50 DO-IT-YOURSELF PROJECTS FOR
KEEPING CHICKENS

Chicken Coops, Brooders, Runs, Swings, Dust Baths, and More!

JANET GARMAN, TIMBER CREEK FARM

Skyhorse Publishing

Skyhorse Publishing books may be purchased in bulk at special discounts for
sales promotion, corporate gifts, fund-raising, or educational purposes. Special
editions can also be created to specifications. For details, contact the Special Sales
Department, Skyhorse Publishing, 307 West 36th Street, 11th Floor, New York,
NY 10018 or info@skyhorsepublishing.com.

Skyhorse® and Skyhorse Publishing® are registered trademarks of Skyhorse
Publishing, Inc.®, a Delaware corporation.

Visit our website at www.skyhorsepublishing.com.

10 9 8 7 6 5 4 3

Library of Congress Cataloging-in-Publication Data is available on file.

Cover design by Abigail Gehring
Cover image courtesy of istockphoto.com

Print ISBN: 978-1-51073-175-2
Ebook ISBN: 978-1-51073-176-9

Printed in China

This book is dedicated to my husband, Gary, my family, and my mom. In the 1970s, my mom went back to work so I could attend the University of Maryland and earn my degree in Animal Science. Since then I've had many jobs. It wasn't until recent years that I have found a way to make a career writing about farm animal care, and have it mesh with our farm and family business lifestyle. Mom mentioned that I was finally doing what I set out to do. Thanks for making the sacrifice, Mom. To my husband, the steady one in our marriage. Thanks for always letting me chase my dreams, start more projects, and bring home more animals. You truly are a life partner in this journey. To my grown-up kids and baby grandchildren, I love you more than I will ever be able to say. I hope you will all follow your dreams, too.

Contents

Preface

by Pam Freeman

Having kept chickens for over a decade, I can tell you the coop and accessories I started out with are no longer around. Over the years, a huge flaw in the placement of my original coop became evident as walking up and down the hill where it was located became tedious, hard to traverse while carrying heavy loads, and downright treacherous in wet or icy weather. An accidental rooster led to a broody hen that hatched more accidental roosters who needed a bachelor pad. A fallen tree left us with no shade and necessitated a porch and trellis to keep my flock cool in summer's heat. Honestly, my chicken coop has completely changed and will continue to change as new needs dictate.

Is this a bad thing? No. Janet is right when she points out that her coops are always evolving. It's a natural part of chicken keeping, and frankly, of life. The key is how you handle this evolution.

No matter whether you build your coop or buy your coop, I've found that everyone's chicken keeping routine and backyard is a little different.

My chickens roam freely for large parts of the day. Some folks keep their chickens confined. My backyard is a hillside, and that needs to be considered when planning improvements. No mobile chicken coops for me! I have a friend that just bought a house and discovered her backyard turns into a river when it rains. Drainage is top of mind for her. Essentially, what works in one backyard may not work in another.

That's where this book comes in handy. Janet details DIY projects for your entire chicken keeping journey, from baby chicks to adult birds. These are useful projects that can be adapted to most any situation and need. If you're looking for a cleaner coop, you can learn how to make a dropping board. If your chickens can't get out to roam, learn how to keep them occupied by making some in-the-coop fun for them. If you'd like to add some healthy treats to your chicken's diet, learn how to make an herb garden out of a pallet.

What I like most about this book is the depth of the projects covered. Some of the projects are flat out fun, but others—like predator prevention and cooling a coop in summer—are crucial to keeping your birds alive and healthy. This is a no-fuss, no-muss book that allows chicken keepers to better the lives of their birds without breaking the bank. This is a book that helps ordinary folks take their chicken keeping to something extraordinary without pretense or unnecessary glamour. It's a book I'm going to keep on my shelf and consult often. I hope you do too!

—Pam Freeman
Author, *Backyard Chickens: Beyond the Basics*
Editor, *Backyard Poultry* magazine
Editor, *Countryside & Small Stock Journal*

Foreword

by Joel Salatin

If I believed in reincarnation as an animal, I would definitely ask to come back as one of Janet Garman's chickens. I always thought that our pastured chickens here at Polyface Farm had the best life in the world, but I think I've found a more pampered—albeit much smaller—flock in Janet's backyard.

Life contains numerous gateway elements. Dr. Seuss is a gateway phonetic and reading introduction to children. Bacon is the gateway meat product for reconstructed vegans. On our farm, we call eggs our gateway product for the many restaurants we serve. If we have eggs, the chefs want chicken, beef, and pork; if we don't have enough eggs, the chef client may skip all the other stuff. A debilitating illness is often the gateway to diet changes and non-chemical, local food sourcing.

For sure, chickens are the gateway to domestic livestock. Whether you're in the country or the city, the day you decide to step beyond the garden and grow your own animal protein, chances are chickens will be the team player of choice. Reasons are many:

- Chickens are small and conducive to small spaces, simple shelters, and lightweight control.
- Chickens are child-friendly; they may scratch or peck a bit, but you won't have to worry about your ten-year-old being trampled by a goat or cow.
- Chickens are cheap—a low investment risk, easy to start, easy to stop.
- Chickens are common; you can get chickens almost anywhere.
- Chickens are quiet—at least if you don't have a rooster—compared to barking dogs, baaing sheep, or mooing cows.
- Chickens have a fast turnover: they become productive quicker than any other barnyard animal besides mice . . . but most of us don't eat mice.

- Chickens are by far and away the superior backyard sanitizer, relishing all sorts of pests, from ticks to slugs.
- Chickens eat kitchen scraps and lay eggs, offering a pathogen barrier between what humans eat (eggs) and what the chicken eats (scraps).
- Chickens are easy to process once their productive life is over; a pot of hot water and a knife are all that's required.
- Chickens are the best role model for teenagers: they're the first animal awake in the morning; they spend all day happily turning trash into treasure; at the first sign of dusk, they head to bed (no gallivanting around the countryside at night).

As ubiquitous as chickens have been in nearly every culture, the fact that in modern America most people have never seen one, petted one, fed one, or enjoyed one speaks volumes about unprecedented collective ignorance. Fortunately, cultural pendulums eventually swing back to correct themselves; the renewed interest in backyard chicken rearing indicates positive changes. Anyone dedicated to integrated environmentalism, personal self-reliance, and bio-regional resilience delights in the backyard chicken tsunami sweeping the country.

One by one, cities are rescinding anti-chicken laws. From high rise apartment complexes to suburban farmettes, chicken keeping is the status symbol of the regenerative elite. Anyone can write a check to an environmental organization. Anyone can patronize local craft beer. But only the sustainability aristocracy become chicken keepers.

In *50 Do-It-Yourself Projects for Keeping Chickens*, Janet Garman takes the wanna-be on a practical and winsome journey to family-stead chicken success. The uncaring-turned-caring food participant need no longer feel intimidated and paralyzed by ignorance. Using household discards and savvy make-do, Janet's projects in this beautifully illustrated manual take novices by the hand, leading them into the incredibly fulfilling joy and satisfaction of chickeneers.

My first flock, 50 as-hatched heavy breed chicks from Sears and Roebuck, arrived by mail in 1967, when I was only ten years old. With a light bulb, cardboard box, and some shavings for a brooder, I still remember my excitement that first morning those many years ago. We had a pet cat. We had a pet dog. But this was different—these were working team members, an entrepreneurial gateway.

Chickens are wonderful partners. They work and sing their little hearts out; they're always happy to see you; and a well-nourished egg, well-laid, is perhaps the most complete food in the world.

While my chicken rearing has always had a bit more commercial and utilitarian, I love Janet's tips for comfort and chicken enjoyment. Simple chicken run features, painted playground equipment, and "boredom busters"—including a chicken jungle gym—offer respectful honor to the chickeness of the chicken.

The love and care Janet exhibits throughout this book is extraordinary: DIY herbal salves, chicken first aid kits . . . you get the feeling that these chickens are part of the family. How delightful in a culture that introduced chicken factories to the world. This is a welcome and necessary paradigm shift.

In my perfect world, even our Polyface Farm would not grow the number of chickens we do. Chickens would inhabit every backyard, every apartment. They would eat all food scraps and be snuggled up next to every commercial kitchen and institutional commissary. Every college dining service would have a hundred chickens next door to close the loop from scrap to human food treasure. This book facilitates such a world, and I hope it'll catalyze timid folks to jump into enriching their lives with a couple of chickens.

If you want a working pet, nothing beats a chicken. So go ahead; jump in. You can create a place of beauty and functionality using this wonderful resource as a guide. Who knows? Maybe you'll provide the tipping point to cascade our culture toward integrated food and functional ecological regeneration. Thank you for caring.

—Joel Salatin
Polyface Farm
Editor, *The Stockman Grass Farmer*

Introduction

We are the new generation of chicken keepers. As we begin raising small flocks of poultry and providing more of the food we eat from our own backyards, we learn to make do. We learn to build emergency habitats for animals. Our sheds fill with construction materials that might come in handy one day. Bring those materials out and put them to good use!

Welcome to the world of DIY projects for chickens. Homeowners, homesteaders, and farmers across the country and around the world enjoy raising backyard chickens. There is something special about these feathered pets. Having chickens to care for, right outside our back doors, brings us back to our roots, adds delicious nutritious food to our tables, and provide hours of entertainment. I'm sure there are people around the globe whose favorite time of day is watching the flock forage for bugs while relaxing after a long day at work.

At some point during chicken raising, we get the urge to upgrade, rebuild, or create a new environment for our feathered friends. Buying all the new materials doesn't always meet with our budget! Some of us prefer a rustic, well-loved, vintage look for the chicken coop. Or maybe you love getting into a new project. Read on! As you will see, projects in these pages range from simple things that take only minutes to complicated weekend endeavors.

Many ready-made coops, accessories, and products are available to the chicken keeper. Often these are out of reach

The author feeds her chickens.
Apron compliments of Fluffy Layers. >>

financially. It can be fun and inspiring to make your own chicken equipment. Even if you have a delightful, prebuilt coop for your flock, adding to their environment is fun. You will enjoy watching the chickens discover new activities, treats, and resting spots.

Trash-to-treasure thinking has become a design trend. Old furniture, recycled containers, vintage rehab projects, and repurposed buildings can all have a part in your chicken area. Even if you have a well-manicured suburban backyard, you can incorporate DIY chicken projects. Using trash doesn't have to be trashy! Spray paint, sandpaper, inspiration, and imagination transform old tossed aside items from the trash heap to chic chicken decor.

Keep your eyes peeled! Garage sales, flea markets, second-hand stores, and landfill centers may be holding your next DIY chicken project! One of our favorite activities, when we can get away from the farm for the day, is attending a tractor pull or steam tractor show. I was amazed to find the best flea markets seem to set up at these shows. Flea markets with lots of farm equipment, vintage furnishings, kitchen tools, dishes, and pans. Was I looking for things for me? Well, sometimes. Mostly I was keeping my eye out for things I could repurpose on the farm. Old tobacco drying frames, discarded beehive frames, and other drying racks made good covers for chicken run gardens. Wooden vegetable bins can become nest boxes, along with vintage wooden crates. Speaking of crates, look for large dog kennels to use for chick brooders or sick chicken isolation coops.

In our minds, DIY is also largely a make-do type of existence. We tend to grab what we have on hand to make the projects. Using what you have and making do with recycled materials is the essence of DIY. Challenge yourself to look around the farm, homestead, or back-yard. Do you have building materials that you can repurpose? The pages of this book bring you many ideas for "do it yourself" inspiration. The measurements and instructions given are largely adaptable to your own

size requirements and materials. Use our directions, but do not be afraid to vary and make each project uniquely your own. Not all projects are right for all people. The style of these projects is rustic, and in many cases, we used what we had on hand at the farm to make them. Raising chickens doesn't need to be costly.

Grab the tool belt and staple gun! It's time to start spoiling your chickens. Make them the envy of all the other chickens on the block. Then sit back and relax, grab a cold drink or a hot cup of coffee, and watch your feathered egg production crew enjoy the good life, DIY style.

Disclaimer: While building any of the structures shown in this book, keep in mind that as we built the examples, some variations had to be made to adapt to our building materials. A basic knowledge of how to build a framed structure will be helpful, although not mandatory. Variations might be seen between what we used and what is shown in the photos or illustrations. We may have had something usable already on site, but the instructions call for a piece more readily available or more cost-effective. This is mentioned because our philosophy is to use what we have available whenever possible. Making substitutions will not affect the finished product but may require slightly different measurements. Please remember to always wear safety eye protection when using tools.

Terms and Materials

I have realized that building materials and tools, as well as chicken coop parts and furnishings, may be called by different names in different parts of the world. Here are some equivalent terms that will reduce confusion.

Chick Brooder—The house or container that the newly hatched chicks are housed in for the first few weeks of life. If the chicks are hatched by a broody hen, the brooder will be bigger than if you are only housing chicks. Alternatively, the broody hen can be left with the flock.

Chicken Coop—The housing provided for the grown chickens. Should be fully enclosed with a pop door and, if applicable, a full-size door. Ventilation and security are key components.

Chicken Wire—Woven product made from thin gauge wire. This product should be used for keeping the chickens in a designated space or keeping them out of an area. It should not be relied on for predator prevention.

Hardware Cloth—Welded wire forming a grid. Available in ¼", ½", and 1" squares for protection and enclosing a space.

Free Range—Poultry roaming freely and eating mostly grasses, weeds, and insects.

Chicken Run—A play and living area for chickens that is attached to the coop.

Water Fount—The two-piece assembly that holds a container of water and the drinking well it drains into.

Feeder—Any system that holds and distributes feed for the chickens to eat. It can take many forms from a gravity-fed system to a shallow pan.

Layer Feed—Grain ration fed to laying hens to provide nutrition. Many laying rations include added calcium. Organic layer feed often does not contain added calcium.

Chick Feed—Smaller particle grain ration for young chicks. Until sixteen weeks of age, no added calcium should be given to chicks.

Grit—Grit is small stone grindings given to chicks and chickens to assist in digestion of seeds, grain, and grass.

Calcium—A mineral that chickens need and can be provided in the form of oyster shells or dried crushed eggshells.

Tools

Most of the projects in this book require only basic building tools. While some photos may show use of a circular saw or a chain saw, a hand saw or hacksaw will also do the job. A power drill/driver will make everything easier.

- Hammer
- Screwdriver
- Tape measure
- T or square
- Level
- Pliers
- Drill bits
- Staple gun and heavy-duty staples

Building Supplies

Just as with the list of tools, the building supplies are basic and easily found in home improvement stores, hardware stores, and home centers.

- Chicken wire (woven, single-ply metal wire)
- Hardware cloth (also called rat wire or welded wire)
- Various sizes of building lumber including 2x4, 2x3, 1x3, plywood, shims
- Hardware for doors and gates, including bolts, latches, hinges
- PVC pipe and fittings
- Cable ties (zip ties)
- Paint (or use what you have left from other projects)
- Various bolts, nails, screws

CHAPTER ONE:
Chick Needs

Your new chicks will arrive soon. What should you have ready for them?

Taking the place of the mother hen requires that we provide all the necessities for the newly hatched chicks.

When not raised by a hen, chicks begin life in a box often referred to as a brooder. The brooder will be their little world for the next few weeks. It will need to hold not only the chicks, with room for growth, but food, water, and a heat source, too. Even with a broody hen, it can be safer and less stressful if the hen and chicks have their own space for the first few weeks.

Brooders can take many different forms. The important point is to house your chicks in a safe, warm, environment. Having a brooder that is easy to set up, clean up, and store are added benefits. Anything from a sturdy cardboard box to a creative homebuilt structure can work as a brooder. Since cardboard is a flammable material though, I don't like to recommend it to people for a brooder setup. You'll need a heat source in your brooder, which can also be a fire hazard. Creating a sturdy brooder environment lowers the fire risk.

Setting up the Brooder

Prepare your brooder before the chicks arrive home. Chicks are very sensitive to temperature changes and chill easily. When you are expecting to pick up chicks from the feed store, or take delivery from the mail carrier, prepare the brooder first.

Check the brooder structure for safety, holes, old litter, or poop, and clean as needed.

Bedding in the Brooder

The chicks' brooder bedding is a confusing choice for many new chicken owners. Some of this confusion exists because as the chicks grow, the bedding options increase. The first few days of a chick's life are precarious. The little guy is trying to find food and water and get used to his spindly legs, all while trying to stay warm and cozy with his new flock members. Giving the chick a sturdy, non-slippery surface is very important during this initial period. For that reason, my recommendation is a rubber shelf liner on the bottom of the brooder and a thin layer of pine shavings. Do not use newspaper during the first week or two.

As the chicks gain muscle control and strength, the brooder can be lined with other substances. Newspaper is an inexpensive option. Not only is it easy to find, but it can go directly into the compost pile with the droppings.

Straw should not be used as bedding for newly hatched chicks. It is a slippery surface and will lead to problems such as spraddle leg. In addition, straw is hard to have in the small brooder environment because of its long strands. There is a product available in some markets called chopped straw. It's a much finer grade of straw, cut into smaller pieces. I think it's an excellent product for nesting boxes. However, it is a bit expensive. I like the product but tend to reserve it for other uses. I will leave that decision up to you. Switching to chopped straw after a few weeks would be a good option for brooder bedding.

Sand is an option that some chicken owners rave about. Again, not recommended for the beginning days of life in the brooder. Honestly, I have never used sand in our coops or brooders. I do personally know chicken owners who do, and swear that it is the best option. Depending on your environment, it can either be a good choice or a wet, soggy, messy, disaster.

Chopped cardboard, dried leaves, and wood chips are other options that some chicken keepers use for bedding once the chicks are at least a few weeks old.

An additional advantage to rubber shelf liner is that it can be washed off, dried, and reused.

Food and Water in the Brooder

Add water and food to the brooder. Most people use the plastic water founts for chicks. They are easy to clean, and when elevated slightly, prevent chicks from getting into the water in the beginning. Placing the water container on top of two bricks, side by side, helps keep the water cleaner by keeping litter from being kicked into the water.

Food can be served in a chick feeder, like a water fount, or from a shallow dish or pie pan. There are drawbacks to using an open container. At first the chicks will tentatively peck at the feed offered. It won't be long until they start scratching around in the dish, kicking the food into the brooder. As the chicks scratch around in the feed, some will poop in the feed. Of course, this is not ideal. A good bit of feed will be wasted, too. The chick feeder prevents them from scratching around in the feed. The chick feeder will hold enough food for a few days during the first week of life for a small flock of chicks. If you have many chicks to feed, there are larger chick feeders available. Some hold up to a gallon container of feed.

The chicken feeders are also compatible with mason canning jars. Buying just the base and using your own glass jars is one way to save a few dollars and do it yourself! Start with a pint jar and you can simply switch to a quart jar when the

chicks need more food for the day. After the first week, you will likely have to fill the jars each day and then twice a day. After a few weeks of age, I often switch to an open feed pan because it holds more feed. But again, assess the feed being scratched out and make the decision that is best for you and your flock.

Grit should be offered to the chicks as early as the second week. Grit is ingested and ends up in the gizzard where it helps grind the dry feed and other substances for digestion. Small bits of sandy gravel can act as grit but they need to be ground small. The easiest method is to buy Chicken Grit products. Even if your chicks are on sandy dirt, I still recommend offering chicken grit.

HEATING THE BROODER

Heating the brooder is essential for the new babies. There are a few ways to accomplish this, but with safety in mind, I recommend only two. Neither is a DIY method. I recommend using either a heat lamp specifically designed for livestock and a red heat bulb, or a shelf-style warmer that sits inside the brooder. There are popular DIY "hacks" described on the Internet. Many of these use standard light bulbs or include other materials not rated as safe for the coop. All I am going to say about these methods is, use at your own risk. A makeshift brooder heater increases the risk of fire.

The heat lamp will need to be securely hung over the brooder so that the heat is directed into the container. Placing the heat lamp directly on the cover of the brooder is not recommended. All of the heat, at that close a distance, can be too hot and literally cook the chicks. Hanging the lamp or suspending it from a secure clamp allows you to regulate the heat directed into the brooder. Raising the lamp higher will cool the brooder down. Dropping the lamp closer to the chicks will increase the heat.

There are thermometers on the market for inside the brooder. My method is much simpler. Observe your chicks. If they are eating, drinking, sleeping, and moving

about the brooder, they are warm and comfortable. Chicks that are chilled will cluster together under the heat lamp. Chicks that are too warm will stay as far away from the heat source as possible, often clustering against a far wall in the brooder. Don't cook the little chicks!

Shelf- or table-style warmers provide a gentle heat that mimics the broody hen's body temperature. This heating device sits on the floor of the brooder and the chicks congregate under the shelf/table to stay warm and go out to eat and drink. It most closely imitates a hen's care.

The specific temperature for newborn chicks is close to 100 degrees Fahrenheit. After the first week of life, begin decreasing the temperature by 5 degrees every few days or each week. When using the heat lamp suspended over the brooder, reducing the heat is done by gradually raising the heat lamp a little bit. Observe the chicks after making any adjustments. If they are exhibiting signs of chilling or overheating, make further adjustments. This is an area where close observation can make all the difference in the health and well-being of the chicks.

Adding to the Brooder Environment

During the first few weeks, gradually add dirt and chopped herbs to strengthen the chicks' immune systems. Good herbs to start chicks on include oregano, thyme, rosemary, mint, and lemon balm. These herbs can be offered fresh or dried. The dirt will slowly add microbes and insects to the chick's diet. In addition to chopped herbs, feeding small amounts of vegetables and

cooked egg are good choices. Chicks can enjoy chopped up peppers, greens, herbs, peas, and squash. Make sure to chop everything into tiny pieces. The caution is to not overfeed treats so that the chicks are too full to eat their balanced ration.

Tiny Roost Bar for the Chicks

This quick project works up in minutes. The small roost bar gives the chicks a head start on learning to roost or perch once they don't need to be under the heat all the time. The baby roost bar shown here was made from scraps leftover from other projects.

MATERIALS AND TOOLS
- 2 scraps of 2x4 for the base ends
- 1 short piece of 1x3 lumber about 10" long, or whatever will fit in your brooder space
- 4 wood screws
- Hand saw, chisel, and a screwdriver

1. Mark where you want the grooves to be, where the roost bar will set into the base pieces. Using the saw and chisel, knock out the two notches.
2. Position the roost bar on the base pieces. Using the wood screws, screw into the base, from the back side, to secure the roost bar.
3. Add a coat of paint, if desired.

Brooder Ideas

Storage Tote Brooder

A brooder made from a plastic storage tote or tub is easy to pull together. Buying as large a tote as you can will last longer than a smaller container. At first the container will seem too large! During the first week or two, cluster the food, water, and heat towards one end of the brooder. As the chicks grow, gradually spread out the food and water. This will give the chicks more room to move about the brooder.

The storage bin I purchased is 106 quarts. The measurements are 33" long by 18" wide and 13" high. You should choose the largest bin you have space for, as that will last you the longest time. In the photos shown, you will see the shelf-style brooder warmer I prefer to use. I also added a photo of the brooder with a heat lamp. The heat lamp would be securely hung from above over the brooder. Raising the lamp a little bit each week will gradually acclimate the chicks to room temperature. Again, close observation is the key.

This is my favorite type of brooder setup for starting a backyard flock of chicks. Usually, newcomers to chicken raising will start with less than a dozen chicks. A large storage bin will be enough room for several weeks.

Wipe out any dust and dirt from the storage bin. Begin setting up the storage-tote brooder before the chicks come home. If you are going to pick up the chicks from a local farm supply store, turn on the heat lamp or brooder warmer before you leave for the store. Add warm water to the drinker. Fill the food dispenser with chick food. Now everything is set to welcome the new chicks home. Use a similar routine if you ordered the chicks from a mail-order hatchery. The hatchery will give you a close estimate of when to expect the chicks' arrival.

Save the lid to the storage tote brooder! The next section shows you how to make a secure cover for your brooder. Snapping the lid in place once you make the appropriate cover will keep the chicks safer and prevent them from flying out (once they figure out how to fly)!

MATERIALS AND TOOLS

- Large storage tote
- X-Acto knife
- Straight edge
- Black marker
- Small bolts, washers, and nuts (suggest ¼" x ⅝" bolts)
- Hardware cloth/welded wire
- Drill and drill bit

1. Using a straight edge, draw lines for cutting out the center of the lid.
2. Using an X-Acto knife or utility cutting tool, slice through the plastic and remove the center piece.

3. Mark the edge where you will drill holes for the screws or bolts.
4. Drill holes for the screws or bolts that will hold the wire on the frame.
5. Cut a piece of ½" welded wire that covers the opening plus overlap so that the bolts can hold the wire to the top frame.
6. Attach the wire cover to the frame using the screws or bolts, washers and nuts. Now you have a very secure top for your chicken brooder.

A.

B.

C.

D.

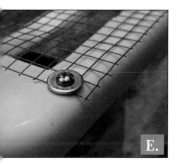

E.

F.

G.

Pet dogs and cats may be curious about your chicks, especially if this is their first encounter with chickens. The tiny chicks' quick movements can look like something the pets want to chase or catch. When you aren't supervising, make sure the brooder is not accessible to the family pets. Even with the wire top on the brooder, the tote is not pet-proof. Using the top may only slow down a dog or cat determined to catch the chicks.

If you're trying to train your dog to be around your chickens without snatching them, here are some suggestions.

Do not attempt to train more than one dog at a time. If you have more than one dog, spend time separately, with each dog, going over the practice exercises. Allow them to meet the chickens only under supervision. Do not just bring your dogs to the chicken coop area and let them run free. Wait until you are confident in the training and behavior.

Remember that instinct will always lie below the surface. Some breeds may not be able to resist chasing chickens and are not to be trusted around chickens at all. It took a long time before we felt comfortable leaving our dogs near the chickens. It may, in fact, take many months or longer.

If you have trained your dog in basic obedience, he will want to please you with his behavior. Reward his good behavior with praise and treats.

If any aggression is shown while the dog is on a leash, end the training session for that day. We don't do a lot of shouting and talking. Training commands are best kept short and firm.

Chicken Wire Brooder

Chicken wire is a useful tool for keeping chicks where you want them. This type of enclosure will corral your new chicks, keeping them safely inside. Chicken wire does not keep out predators. Even your family dog or cat can break into a chicken wire enclosure. Be careful when using chicken wire enclosures for your chick brooder.

That does not mean chicken wire can't be a successful choice. A chicken wire brooder would work well inside a garage, basement, shed, or inside a larger chicken

coop. These are easy-to-construct brooders. Just be aware of the inherent danger in order to keep your chicks safe.

Chicken wire brooders are easily expandable. They can be set inside a child's swimming pool for easy cleanup. Add pool noodles around the top edge for added stability. Cover with screen or chicken wire to keep the chickens in, once they begin to try flying.

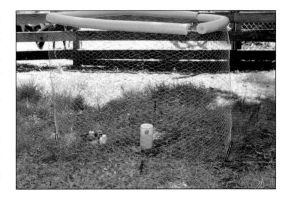

Note: Although the photo shows this chicken wire brooder sitting on grass, this is only recommended as an indoor brooder enclosure. It could be used inside a larger chicken coop to separate the chicks from the flock, or kept inside a shed or garage that closes securely. If you use this type of brooder outside, you would need to supervise, so that predators don't attack and eat your new chicks!

MATERIALS AND TOOLS

- Chicken wire
- Pool noodles
- Wire cutters
- Sharp knife

1. Slice each pool noodle open on one side. Place the opening over the top rail of the chicken wire pen. This may require another person to hold the wire steady.
2. Use metal pins or tent stakes to keep the brooder in place.

REMEMBER! This pen is not a safe brooder if predators or pets can access the area. This works well as a temporary pen while your older chicks are acclimating to the coop and existing flock. It would be fine in an enclosed garage or shed.

A.

B.

C.

D.

Water Trough Brooder with Security Cover

Here is reason number one for saving a leaky water trough from the livestock field: It can be repurposed as a brooder that is easy to set up and easy to clean! I recommend making a simple wooden frame with welded wire as a top over the water trough. The chicks will try out their new wings sooner than you think. It's good to have a top ready to place over the brooder to keep them in place. A wire lid also protects the chicks from family pets.

A.

B.

MATERIALS AND TOOLS
- Water trough
- Wood for the frame
- Welded wire or chicken wire
- Hammer, nails, saw, wire cutters, staple gun

1. Measure the top area of the water trough. Add an inch or two to both the length and width. This will make the top big enough to sit on the edge of the water trough.
2. Cut the lumber. I suggest using 1x3 boards for this project, although any framing lumber will work.
3. Assemble the frame.
4. Measure the area of the frame that will be covered by wire. Make sure the wire will extend far enough into the frame to be secure and sturdy. Cut the wire using the wire cutter.
5. Using small nails or a staple gun, attach the wire to the frame.
6. Place the frame on the water trough.

C.

D.

Pallet-Style Brooder

Building with pallets is a popular trend in repurposing materials. With as little as four pallets you can build a 4'x4' brooder for your chicks. Using more pallets, the brooder can be adjusted for a larger flock.

Find four pallets that are in good condition. Pallets are available from many stores, garden centers, and feed stores, to name just a few. It's always good to ask the store if you can have some pallets before assuming they are trash. In my area, I have never been told no, when asking for pallets. Most pallets are not treated with chemicals as they once were. The newer method is heat treating. Look for the stamp "HT" on the pallet so you know you have a newer, non-chemically treated pallet.

TOOLS AND MATERIALS

- 4 pallets
- 2 pieces of 2x4x8 lumber
- 2 4x4x8 posts
- Hardware cloth (if being used outside) or chicken wire (if being used inside)
- Screws, washers, nails
- C-clamp
- Hammer
- Drill and drill bits
- Saw

1. Cut each pallet down to three-foot height. This is an optional step. I don't like leaning over the full four-foot height of the pallets to grab a chick from the brooder. We used a chain saw to cut the pallets to the size I wanted. A handsaw should work for this, too.

2. For each section of pallet, cut wire to cover the pallet on the inside. This will keep the chicks from escaping through the pallet boards, and, if you use this outside, keep predators from breaking into the brooder. Attach the wire to each pallet. Use hardware cloth wire if the brooder will be left outside. If this brooder will be used inside a garage or shed, you can save some money by using chicken wire.

3. Stand two sides up and attach a post to hold them together at the corner. (Use a cinder block to brace one side in place while working.) Use a drill and a long

drill bit to make the holes for the bolts. Insert the bolts into each corner post. Continue to construct the box by adding the corner posts to the frames. (You could also choose to eliminate the corner posts and directly connect the four pallet sides to each other at the corners. We felt this was less sturdy and opted to add the posts, but it could be done.)

D.

4. If you will use the brooder outside, you will need to put sturdy wire on the bottom so that predators cannot dig under to reach your chicks. It takes a determined fox only seconds to dig under a structure. Cut a piece of welded wire larger than 4'x4'. You may have to use two lengths that slightly overlap in the center. Set the brooder on the wire. Bring the excess up and attach it to the sides of the brooder. Next step will be making a top to keep your chicks in the brooder and other animals out!

E.

5. Measure the top and cut lumber to make a simple frame. Our top shows some overhang. If you prefer you can make a top that fits more securely to the brooder. Again, wire choice will depend on where you will use the brooder. Attach the wire to the frame. We used a staple gun for this. Small finishing nails would also work. Add extra support for the top if you feel it is necessary.

F.

G.

How do I know if a pallet is safe to use?

Look for the code stamped on the pallet. Most pallets are now treated with heat rather than toxic chemicals. If the pallet is stamped "HT" it is safe from chemical treatment.

Does using pallet lumber really save you money?

What if you don't have access to free pallets? The cost difference between using pallets for the sides of the brooder and using lumber is not great. While it's nice to have the sides almost built, when using pallets, the fact is, a few extra steps and a few more dollars are the only differences.

Pallet Brooder
4 free pallets
(2) 4"x4"x8' posts @ $8 each = $16
1 roll of hardware cloth wire = $50
16 bolts @ $1.70 each = $27.20
Total = $93.20

Lumber Brooder
(8) 2"x4"x8' boards, cut in half to make
    the (16) 2x4x4 side boards $30
(2) 4"x4"x8' posts @ $8 each = $16
Wire = $50
Bolts = $27.20
Total = $123.20

Work Bench or Table Brooder

There was wasted space under the work bench in our feed shed. Once cleared out, it made a perfect space for a chick brooder! This same procedure could be adapted to a repurposed, sturdy, table. Frame with lumber, add wire, and you have an almost instant brooder!

MATERIALS AND TOOLS
- Framing lumber (2x4 or 2x3 lumber purchased at a local home center)
- Door hinges and hardware
- Nails
- Hammer
- Saw
- ½" hardware cloth, welded wire, or wire of your choice (also available in home supply centers and farm and garden stores)

1. Clear out any accumulated stuff and trash, then frame space using 2x4 lumber cut to size. The sizing will differ based on the table or workbench area you are framing. Leave a space for a door into the structure.
2. Attach ½" hardware cloth wire to the frame.
3. Frame and build a door to the brooder. Attach hardware cloth and hinges. Attach to the frame.
4. Add shavings, food, water, and grit. Finish by hooking up a heat source for chicks under eight weeks of age.

Using the Brooder after the Chicks Move on to the Coop

After the chicks have grown and are integrated with the flock, hold on to the brooder. Some of the brooders described here can be used for other purposes. I save our brooder setups for sick or injured chickens that need to be quarantined. In addition, any new arrivals that require a quarantine period before joining the flock, broody hens and their nest of eggs, or a rooster who needs a time out, can visit the brooder coop.

Quarantine time is extremely important to keeping your flock disease-free. Even a hen from your neighbor's flock has the potential to bring your flock a serious illness. Signs and symptoms of illness can be non-existent or not showing. By the time you see signs of illness in a new chicken, you have exposed your flock to the illness. Take the time to quarantine. Waiting ten days to a month is a small price to pay for peace of mind and flock health.

Did you know that often a bully rooster or hen can be cured of their bad temper by removing them from the flock for one to three days? This is a great use for your brooder pen. The bully is removed to give everyone else some breathing space. While the bully is gone, a new pecking order is also established. When the bully is returned, he will have to find his place in the pecking order again. This very often works to calm a mean hen or cranky rooster.

CHAPTER TWO:
Intro to Chicken Coops

Big chicken palaces, tiny urban coops, and small humble chicken tractors all provide protection for the flock. In addition, all coops have similar accessories in common. Nest boxes, roost bars, dropping collection, and even food and water areas are needs to be accommodated.

When building or renovating a structure into a coop, take each of these needs into consideration when planning the layout. Many times, you can adapt the larger items to a smaller scale for tiny coops. Personally, I love reusing old items for the coop needs. Finding old crates, chicken feeders, and ladders that I can add to my coop, adds to the fun.

Building a large chicken coop from scratch is a big project! The advantage is that it can be exactly what you want it to be, and contain all of the necessary interior accessories.

Try renovating a garden shed into a chicken coop. This is one of the renovation floor plans we have used in the space. This coop provides plenty of room for our large flock of

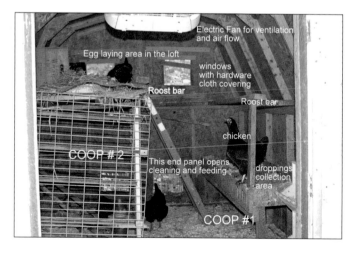

standard egg layers. It's always good to allow at least 2 to 4 square feet of space per bird inside the coop. If your chickens will need to be enclosed in the coop for longer periods of time during the day, increase that space need to 7 or 8 square feet per bird.

This is the before picture of the garden shed, which we turned into a chicken coop. It is a large shed, probably larger than a coop needs to be, but I managed to fill it with chickens. Windows were cut into the back wall and covered with 1-inch welded wire to keep out predators. The same was done to the front barn doors. On the side, a small pop door was added for chicken use. During the winter, only opening the pop door conserves heat inside the coop. Rectangles of plexiglass are installed to cover the windows during the colder months.

Cement or concrete on the floor prevents rodents from chewing into the coop from below. In addition, building your coop 6" above ground level helps prevent this.

The hardware cloth was put on the inside of the coop window. Strips of lumber were used to firmly attach the wire to the walls and doors, using screws. Small pieces of plexiglass are slid in between the strips of wood and the wire during the winter. Using this method, I can easily open or close off a window if the weather demands more or less ventilation.

The floor joists from a coop building project provide a temporary playground for the chickens.

Points to Remember When Renovating or Building

The coop will require the same elements no matter if you build a large poultry palace or a small chicken cottage. Doors, windows, ventilation, window coverings, nest boxes, roost bars, dropping boards, and possibly food and water areas need to be addressed.

Should you keep food and water in the coop?

In a perfect situation, I would say no. However, none of us live with a perfect setup. Some of us work shift work. Some have long days at work. We all want our chickens to have a good life and to be safe. If you occasionally must keep the flock in due to weather or for short term safety reasons, then of course you will need to provide food and water inside. Chickens don't need food or water after they go to roost until they get up around sunrise. If you cannot meet that demand then you will need to have some food and water in the coop. Be aware though that doing so will attract rodents. In addition, the coop will need to be cleaned often to remove spilled food.

COOP VENTILATION

Ventilation is necessary for the air quality of the chicken coop. If you build a coop from scratch, make sure you plan vents along the top of the walls for proper ventilation. These should be wire covered to keep out predators. Prebuilt coops should have ventilation built in. One of our purchased coops had ventilation near the roof, but it proved inadequate

and the coop was always hot and stuffy, even with the door open. We added a second wire-covered opening on the wall near the roof and it solved the problem.

Tractors and Moveable Coops

A movable shelter, capable of housing a flock of chickens while giving them access to fresh grass and insects, is called a chicken tractor. Tractors are a form of chicken protection and housing. This type of coop or housing can be ideal when building a permanent structure is not an option.

Moving the chicken tractor daily or a few times a week provides fresh food for the flock without resorting to purchased grain or commercial chicken food. You must continue to provide a calcium supplement to ensure that the fresh eggs have strong shells. Water must still be provided, too. The chicken's food will consist of fresh green grass, weeds, and insects.

Structurally, the chicken tractor should be fully enclosed with welded wire or hardware cloth wire, including the bottom. This will add a higher level of protection for the chickens in the tractor. In addition, securing the tractor to the ground with stakes will prohibit larger predators from pushing the tractor over.

Just as with permanent chicken coops, the chicken tractors can be made in various sizes and styles. Keep in mind that unless you have a farm tractor, you will be pulling the chicken tractor manually to the new locations frequently. Building the chicken tractor too heavy can make moving it an unwelcome chore.

Types and Styles of Chicken Tractors

Movable chicken structures have been in use for generations. Any lightweight enclosure that can be moved from one spot to another as chickens graze an area is considered a tractor. Sturdy tractors can be made from lumber or repurposed dog houses, play houses, and small sheds on wheels. Adding an enclosed wire area keeps the chickens safe from flying predators. These are particularly useful when you have large pastures to feed your chickens. In addition, a chicken tractor is ideal for backyards, to keep the chickens from digging up one section of the yard. Move the tractor as often as necessary to provide fresh grass before an area has been grazed and scratched down to dirt.

Lumber-Framed Rectangle-Style Chicken Tractor with Attached Coop

While building this chicken tractor or any of the structures shown in this book, keep in mind that as we built this model, some variations had to be made to adapt to our building materials. A basic knowledge of how to build a framed structure will be helpful although not mandatory. This is mentioned because our philosophy is to use what we have available whenever possible. Making substitutions will not affect the finished product but may require slightly different measurements.

In designing this chicken coop, we tried to keep the measurements to two-foot and four-foot increments to save lumber and reduce waste. The finished dimensions of the tractor are 4'x2'x8'. It comfortably houses two to three full-size chickens if the tractor is moved to fresh pasture frequently.

MATERIALS AND TOOLS

- 2x3 lumber (we used (7) 8' boards)
- 1x3 lumber (we used (4) 8' boards)
- 3 sheets of 2'x4' plywood
- Tin roofing (we used a scrap piece from another building project)
- Nails
- Screws
- Hinges and latches for the doors
- Skids or wheels, if desired.
- 2' and 3' wide rolls of ½" welded wire (hardware cloth)
- Hammer
- Drill
- Saw

1. Prepare a work surface.
2. Place two 2x3 pieces of lumber down and attach the 2x3x24" cross braces. Back two cross braces are 24" apart, forming the base for the coop part of the structure. The upright pieces are 2x3x24" and are attached to the base at 24" spacing.

A.

B.

C.

D.

3. Use 24" square plywood for lower base of coop section. Cut out corner pieces 2x3".

4. Attach support pieces for upper floor of coop, using 1x3x24 or lumber of your choice. Do the same for front and back of the coop section.

Make a second plywood floor and cut out notches for the supports. Attach the sides using two pieces of 2'x3' plywood. The door on the back can be attached later in the process.

E.

F.

5. Measure the front opening and build a small door for access to the front part of the tractor. Cover the door frame with hardware cloth. Attach with hinges and a latch.

6. Top railing pieces are attached to the supports to add stability and provide a place to attach the wire. To make a sloped roofline, attach a 2x3x24 to the front edge of the coop section. Attach it on a 2" side.

7. To enable water to run off the back of the tractor, attach a 1x3x24" cross piece flat to the back of the top, as shown.

8. Measure and cut the hardware cloth wire for the bottom of the tractor. Move the tractor onto the wire, and arrange evenly. There will be enough overhang of wire to fold up and attach to the bottom supports.

9. Hardware cloth can be attached with nails partially hammered in and then bent over, or heavy duty staples and a staple gun.

10. Measure and cut the top piece of wire from the three-foot tall wire roll. Attach wire from the coop section to the end of the tractor. Attach the wire to the frame. Measure and cut the

wire for the two sides using the two-foot tall roll of hardware cloth. Attach to the frame.

11. Measure and cut a 2' square piece of wire for the roof of the coop part. This will be under the tin roof, allowing for ventilation and predator prevention.

12. Using a 2x4' piece of plywood for the door, attach to the back of the coop with hinges of your choosing. On the other side of the door, install a sturdy latch for closing the coop. Attach a piece of tin roofing, allowing enough for overhang.

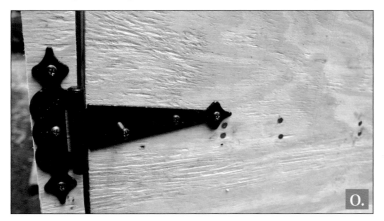

13. When the structure is completed, give the tractor a coat of paint to help preserve the wood. Add a roost bar to the coop section.

Other Chicken Tractor Ideas

Photo by Homestead Honey, https://homestead-honey.com

Photo by Homestead Honey, https://homestead-honey.com

Fixer-Upper Chicken Coop

Almost any pre-existing structure on your property has the potential to house chickens. We had an old rabbit hutch that wasn't being used. While not large enough for standard egg layers—at least not more than two of them—it was large enough for a small flock of bantam chickens. The small structure needed some adaptations to make it right for chickens. Since I was already deep in the DIY territory, the coop underwent a big cosmetic rehab, going from drab to chic!

First, a roost bar was added. This was simply a piece of 1"x2" trim lumber that was inserted through the hardware cloth in the upper section. During the winter months this open section will be fitted with plexiglass "windows" to block the cold air.

A section of pallet was brought in to make a front porch and the entire area was fenced using a chain-link fence package. The upper portion was fully enclosed using chicken wire, keeping the chickens in the pen and aerial predators out.

A scrap piece of lattice was used to add some shade to the front porch.

Never underestimate the decorating power of a can of paint.

The shutters were discovered at a local salvage shop, five dollars for three shutters. Now to find a project for the third shutter!

E.

As a final touch, I added a porch swing. The cost of this transformation was budget-friendly. Consider what you can do with an unused dog house, children's play house, or garden potting shed.

PVC Chicken Tractor

A cheaper, lighter version of a chicken tractor can be built using PVC plumbing pipes and chicken wire. The important caution with this plan is that it doesn't offer terrific protection from predators. Chicken wire is not a strong deterrent to most predators. Upgrading the wire to a hardware cloth would add safety, but the structure is still lightweight.

Ends of a wire clothes hanger can be repurposed as stakes to anchor the tractor to the ground.

PROS
- Easy to assemble
- Easy to break down and store for later
- Easily moved from site to site
- Costs less than twenty dollars

CONS
- Houses a small number of chickens per tractor
- Chicken wire is not sturdy against predators
- Easily tipped over by other animals and pets, possibly harming chickens

A PVC-style chicken tractor can be successfully used inside a larger chicken run when integrating new chickens to the existing flock. Adding the PVC chicken tractor to the front of a permanent chicken coop can act as a small, covered chicken run.

MATERIALS AND TOOLS

- 2 PVC pipes (¾"x10') (You will need three sections at 5', two sections at 2' and four sections at 1'. That is the total for PVC pipe. There will be a small piece left over.)
- Various 3-way PVC pipe fittings for the connections as follows:
    - 6 PVC right angle elbows for ¾" pipe
    - 2 PVC tee connectors for ¾" pipe
- Approximately 20' of chicken wire
- Assorted sizes of zip ties (cable ties)
- Hacksaw or handsaw with fine teeth
- Wire cutters for cutting the chicken wire

1. Cut the long PVC pipe into two 5' lengths of pipe. Make the base of the PVC pipe tractor by laying out the two 5' lengths of pipe approximately 2 feet apart.
2. From a second pipe, cut four 1' pipe sections.
   Using the tee connectors, lay out the front and back of the bottom frame. Then connect to the 5' sections using the elbow connectors.
3. Lay out a 5'5" piece of chicken wire, allowing for excess on each end. Place the base frame on top of the chicken wire and use cable ties to tightly attach the chicken wire to the PVC frame.
4. Attach the two 2' long PVC pipes to each tee connection from the base. This forms the uprights at each end.

5. Using a 5' length of pipe, attach the top bar to the uprights using elbow connectors.
6. Cut a piece of chicken wire approximately 15' long. Loosely wrap around the chicken tractor frame.
7. Begin to attach the chicken wire to the frame using the cable ties. It will be floppy at first. Keep adjusting and bending the chicken wire to attach it to the top bar. Adjust the ends as needed. Keep the wire straight at the bottom so you can connect to the bottom frame.
8. Connect the chicken wire to the base frame and the base chicken wire.

I did not make a door for this coop. The chicken wire opens easily at the connection and I simply bend a few of the wires over to close it.

Caution

This will keep a few chickens happily grazing and is lightweight enough to move easily. It is *not* predator-proof. Do not leave chickens in a chicken wire enclosure where wildlife, dogs, or other predatory animals can gain access to them.

A Note on Chicken Grazing

Tall grasses can cause problems with impacted crop in chickens (see page 53). It is advised to cut grass to a few inches in height before letting the chickens graze. Often farms will let larger animals such as cattle or sheep graze an area first. The chickens follow and clean up the pasture. You may not have large grazing animals to help you with this. Use caution and cut the grass with a lawnmower before putting the chickens out in the tractor.

Dropping Boards, Dropping Areas, Nest Boxes, Roost Bars, and Perches

Let me tell you a humbling story. When we planned our first chicken coop renovation, I never even considered the basic needs of chickens in a coop. We didn't research what the coop should have in it. We just had a shed and decided to use that for a coop. Wow, did we have a lot to learn.

Chickens need a perch or roost bar set up for sleeping. I can't imagine what those first chickens thought of our plan! There were a few structures in the coop but not a well-built perch. I think they slept in the row of nest boxes, which is probably why they didn't lay eggs in the nest box.

Thankfully it wasn't long before we realized that the chickens weren't supposed to sleep in the nest boxes and that a roost bar would need to be added to the coop. A long 2x4 was added to the coop, directly over the nest boxes. Where do you think all the droppings from the chickens landed? Yes, right in the nest boxes!

It wasn't long before I realized another important feature of the coop had been left out of our design; nest boxes. Instead of open nest boxes under the roost bar, I added wooden crates, set on their sides, making enclosed nest boxes. Many of these crates were vintage finds from flea markets. Really any sturdy container will work, as long as it can be securely set in place so it won't tip over.

Over top of the wooden crates I placed a long, wide, dropping board. Empty feed bags are used to cover the dropping board and make daily cleaning quick and easy. To clean the dropping board, I grab a dustpan and kitty litter scoop. I scrape the droppings onto the dustpan and toss it all on the compost pile. It takes only a minute to clean up the overnight droppings in the coop. Now the nest boxes stay clean for egg laying.

Doing the daily cleanup controls the fly population, since leaving the droppings in the coop attracts more flies. Keeping the coop clean of droppings means the eggs stay cleaner, too. Chickens have no problem walking through droppings on the way to the nest box. Those dirty feet then step on any eggs in the nest.

This is one method of adding a dropping collection area when converting a shed into a chicken coop. Any area that is easy to clean under the roost bar will work. Some coops have a tray set up. The tray slides out for easy cleaning. Another idea is a hammock made from a tarp and hung under the roost bar. Detaching the tarp and dragging it from the coop to the compost pile and then putting the tarp back is all it takes to clean up the droppings.

When you build a coop from scratch, be sure to include nest box areas, perches, roost bars, and a dropping board or dropping collection area.

This photo shows a cleanup method using layers of washable tarps. Each day the top tarp is removed from the coop. The droppings are added to the compost pile and the tarp is hosed off and hung to dry. Photo by Brittany May, www.happy-days-farm.com

Photo by Brittany May, www.happy-days-farm.com.

This coop has a tray that pulls out to clean the dropping area. Photo by Brittany May, www.happy-days-farm.com.

Nest Boxes

Some common items used for nest boxes include plastic bins, laundry baskets, cat litter boxes, buckets, and wooden crates. Plastic containers are easy to clean and inexpensive to replace. However, plastic containers are lightweight and tip over easily. Refer to the section on building a frame for nest boxes (page 38) to make sure they sit securely in the coop.

Wooden crates look rustic and are comfortable for the hens. I set ours on the side so the nest is enclosed on three sides. Adding nest box curtains gives the hen even more privacy during laying. Adding a secured plastic bin to the inside of the wooden box makes cleanup extremely easy. Painting the inside of the wooden nest box will help control wood mites.

Diatomaceous Earth powder (always use the food-grade DE powder) and herbs in the nesting material help to deter pests. A homemade non-toxic cleaner made from vinegar and essential oils (see page 113) or fresh herbs also keeps the coop and nest boxes fresh.

Plastic crates, wooden crates, plastic bins, and laundry baskets all work as nest boxes. Other options include empty buckets, cat litter pans, and large flowerpots.

Straw and pine shavings are good choices for nesting material. I like to use a mixture of the two.

Remember, don't position the opening of the nest boxes under the roost bars!

A Frame for Nest Boxes

Plastic tubs or lightweight boxes that aren't secured to something can tip over, trapping the chicken under the plastic bin, box, or bucket. Construct a frame that holds the nest boxes in place. Simple frames can easily be built from lumber. At the very least make some way to fasten the nest boxes in place to avoid tipping. This frame will hold three 12"x18" nest boxes.

MATERIALS AND TOOLS

- Two 14"x65" boards
- 1"x2"x65" board for the front lip or edge
- Three 1"x3"x65" boards for back of the structure
- Two 1"x14"x18" boards
- Nails or screws for assembly
- Handsaw
- Drill or hammer
- Nest boxes (painted to repel insects)
- Cinder blocks

1. In order to have the top slant forward, you need to cut the side pieces at an angle along the top edge. For each end piece, cut the top diagonally from the top back corner, which is 18", to the front, making the front 14". Screw or nail the side pieces to the outside edge of the bottom board. Add the board for the top.

2. Nail the front edge to the front of the bottom board, extending the length of the front.

3. Nail the three boards along the back of the nest box frame.
4. Insert the three nest boxes. Place the nest box frame on top of two cinder blocks or some other elevation.

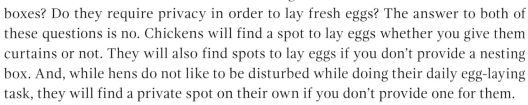

Nesting Box Curtains

Do chickens need curtains on their nesting boxes? Do they require privacy in order to lay fresh eggs? The answer to both of these questions is no. Chickens will find a spot to lay eggs whether you give them curtains or not. They will also find spots to lay eggs if you don't provide a nesting box. And, while hens do not like to be disturbed while doing their daily egg-laying task, they will find a private spot on their own if you don't provide one for them.

Having nesting boxes, and even adding privacy curtains, encourages the hens to lay eggs where we can conveniently find them. Rather than searching the barn, haystacks, and surrounding landscape to find the daily dozen, the eggs are right there in the coop where we can collect them easily.

Hens that have a quiet, secluded environment to lay eggs suffer fewer side effects from stress and are more productive. Providing a safe space for the hens is beneficial to their overall health and lifespan.

To make these nesting box curtains, I used a piece from a sofa slipcover that was heading to the trash. The skirt from the slipcover was the right depth for the curtains. This made the job easy because there were no raw edges to hem. You could also use an old bed skirt, or any piece of fabric that you have or like.

I began by stretching the fabric out the length of the nest boxes. I chose to use a staple gun to fasten the fabric in place. If you have smooth wood surfaces, a hot glue gun may work just as well.

After securing the top edge of the fabric to the top edge of the nest boxes, I cut slits where I wanted the openings for the chickens to be. Using more scrap fabric, I made some tie backs to open the nest box curtains slightly.

Roosting Bar

Roosting is what chickens do when they sleep at night. If the chickens didn't have a coop, they would fly up into trees or rafters in the barn and find a comfortable bar to roost upon. When chickens sleep they go into a very deep trance. It's a wonder they don't fall from the roost!

Having a roost available in the coop provides a feeling of security for the chickens. This is what they would do in nature. If you don't have roost space, they will find a way to roost in the nest boxes or on top of anything available.

When building the roost, allow approximately 8 inches of roosting space per bird. If you have big heavy breed chickens, allow a bit more space. From the time the chicks are little, they will begin to try to roost. It's natural and will be learned by chicks, even if they're not raised by a broody hen. The tiny roost bar for chicks should be only slightly above the floor. As the chicks grow, raise the bar. The roost bar should be positioned higher than the nest boxes or, again, the birds will try to roost in the boxes. This is due partly to them wanting to roost on the highest points available, as they would in nature. I have a few that will work their way up to the rafters! While I feel this is a bit too high to be encouraged, they persist. If you find your flock choosing really high roosting bars, make sure that there are intermediate levels for when they get off the roost. Chickens don't fly very well and you do not want them coming to a hard landing from a roost that is too high.

Building roosting bars should be one of the simplest projects to add to the coop. A simple 2x4 piece of lumber makes a comfortable roost. Make sure to securely install the roost bar so it is stable. Chickens won't choose a roost that is unstable. In addition, the 2x4 gives the chicken feet a stable solid "footing." Smaller round roosts require the foot to clutch on harder. Smaller perches are best left to the yard, where they are temporary rest stops. Plastic and PVC pipes are slippery and the chicken's foot must grip even harder to stay on the roost. Do not use slippery plastic tubes or pipes or metal. Metal will get cold and possibly allow frostbite on the feet.

Whenever possible, keep your chickens from sleeping on the floor. If a chicken is injured and cannot roost, place it in a crate on a deep bed of straw for warmth and a secure atmosphere. In addition, the floor of the coop is where the poop lands. It's likely to have insects and pathogens. Sleeping on the floor makes your bird more likely to become ill.

A simple 2x4 attached to the coop wall is probably all you need. For more chickens, a traditional roosting structure will serve the purpose of getting all the chickens to roost at night. This structure is easy to build from framing lumber. It can be modified and built from tree limbs too. Make sure to use sturdy limbs that are thick enough for the chickens' feet to rest comfortably on the roost.

Painting the lumber with a coat of paint will help discourage wood mites.

The following structure can be adapted to smaller size coops. The general sizing guideline places the bars 10 to 12" apart for standard chickens and 12 to 18" apart for large breeds. The levels are angled out at a 30 degree or greater slant to prevent the chickens on the upper roost bars from pooping on the backs of the lower roosting chickens. We've all heard the saying about being higher on the pecking order!

MATERIALS AND TOOLS
- Lumber
- Handsaw (or circular saw, if you prefer)

- Nails and screws
- Tape measure

1. Measure the length of the lumber for the roosting bar and cut the lumber. You will need three of these, one for each end and one in the center.
2. Measure the length of the lumber needed for the back. This piece will be attached to the wall of the coop.
3. Attach the three support legs coming out from the back. Space the legs evenly.
4. Measure for the actual roost bars. These pieces will extend the width of the roost and will be the length of the back board.
5. Attach the roost bars to the support legs using nails or screws. The roost boards will sit at an angle from the support legs.
6. Paint the roost. Make sure all the chickens can be outside while the paint dries. Also, make sure that there is plenty of ventilation. Use a fan to help disperse the paint smell.

Painting the roost keeps pests like mites and lice from burrowing into the wood, and then later attacking your chicken's feet.

Outdoor Roosts and Resting Perches

Chickens appreciate a place to rest during the day. Placing outdoor perches and roosts around the chicken run or in the yard lets them rest up off the ground.

Our simple solution was to secure some sturdy branches to the tops of tree stumps. This gave the flock a selection of perching spots to choose from. The logs hid insects for them to find, as an added bonus!

This type of roost could also be used in the coop. Make sure the branches are strong and large enough that the chicken's feet are comfortable while perching. Don't use thin branches.

Feeders and Waterers

Food and water can be served to the flock using a variety of containers, many of which can be homemade, or fashioned from recycled parts.

PVC Pipe Feeder

Make a grain feeder from any size PVC pipe and fittings. These are so easy to make and can be spruced up with a simple coat of paint.

As the weather turns cooler, rodents will be more persistent at trying to get into the coop for food. Any food spilled on the ground will attract rodents and soon the whole rodent population will be dropping by for a snack before bed. Mice and rats don't need a very large hole to gain entry. Before you know it, they'll be eating a significant amount of that pricey layer ration you bought for your chickens.

The best response to a rodent problem is to avoid leaving any food in the coop. If you must leave food in the coop due to predators or your daily schedule, try these PVC feeders. They significantly cut down on spilled food. Adding the caps to the top and bottom will stop rodents from gaining access to the food while the chickens are roosting. Simply remove the plug in the feeder end when the chickens get up in the morning.

MATERIALS AND TOOLS
- PVC pipe and fittings
- Spray paint (optional)
- 2 PVC end caps

- Glue gun
- Hose clamp

1. Spray paint the pieces of pipe and fittings, if desired.
2. Glue one round end cap into the bottom of the pipe. This will keep the grain from falling out of the feeder at the bottom of the tube. I used a hot glue gun to do this step. If you have it on hand, you could use the pipe fitting cement sold at home improvement stores.
3. To install the feeder, use a hose clamp, or some other assembly to attach the pipe feeder to the wall or post. When it is securely in place, fill the pipe with the grain feed.
4. Use the second end cap to close off access to the feed. Put the cap on the top of the pipe after filling.

Filling the feeder with only enough feed for two or three days should prevent moisture and mold from being a concern, but check periodically. If you have a large flock of chickens you will need more than one PVC pipe feeder. My recommendation would be one feeder for every three or four chickens. This will ensure that everyone gets plenty to eat.

This is a good solution to very limited floor space in the coop. The PVC pipe feeders take up very little space and prevent feed from being scratched out and wasted. It is also almost impossible for the chickens to poop in these feeders!

The PVC pipe feeder can also be used to hold supplements such as oyster shell for calcium. Photo by Cheryl Aker Hubbard, www.pasturedeficitdisorder.com.

Using a hook and metal strap, the feeder tube can be easily detached and removed for the night to prevent rodents from having a feast. Photo by Cheryl Aker Hubbard, www.pasturedeficitdisorder.com.

Photos by Cheryl Aker Hubbard, www.pasture deficitdisorder.com.

Bucket Feeder or Waterer

MATERIALS

- Plastic buckets and lids marked as food safe. (Look for the recycling symbol on the plastic. The number 2 means it is food-grade plastic.)
- Drill

1. Drill 6–8½" holes in the bucket 2" from the bottom.
2. Place the bucket in the tray that will hold the water. Fill the bucket with water.
3. Immediately put the locking lid on the bucket to stop the water from running out completely. Attaching the lid with the seal is the key to this waterer working as described. In addition, make sure the bucket is set on level ground or a level platform.
4. Use the same method to make an automatic feed dispensing bucket feeder. Instead of using many small

holes, make the holes larger and drill only two to four holes for the feed to come out of the bucket into the tray.

Variation on Bucket Waterer

- One plastic bucket large enough for the jug to fit inside
- One 5-gallon water jug
- Knife or scissors

1. Cut out a hole near the bottom of the bucket, large enough for the chicken's head to fit through.
2. Fill the jug with water. Turn upside down in the bucket. Only enough water will flow out to fill the bottom of the bucket.

Storage Tote Feeder

MATERIALS AND TOOLS

- 1 plastic storage tote with locking lid or large plastic bucket with lid
- 4–6 PVC fittings in shape shown
- Drill with large hole bit
- 1 tube silicone sealant

1. Drill holes in the sides of the storage tote, 2" from the bottom. Space the holes evenly around the storage box.

2. Insert the PVC fittings so that the feed will not pour out of the openings. Fill with feed, cover with the lid.

Rain Gutter Feeder Trough

Nothing fancy is needed for this project—just a scrap of an old rain gutter, or a cheap one from your local home improvement center (they should only cost a few dollars), some scrap lumber, and a rope or chain. If you have a building supply re-seller or charity store in your area, you can probably find everything you need there.

MATERIALS AND TOOLS
- Rain gutter
- Gutter end caps (optional)
- Scrap lumber pieces
- Rope or chain to hang the feeder
- Measuring tape
- Drill

1. Measure how long you want the feeder. Think about where it will hang and what type of clearance is needed at each end.

2. If you will be using the feeder outside, drill tiny holes in the bottom for drainage in case you have a storm while feed is in the feeder. You can buy end caps for the gutters for a couple of dollars more.

3. At each end, drill a hole large enough for a rope or chain to hang the feeder off the ground. Hanging the feeder will save feed by preventing the chickens from

scattering the grain with their feet or pooping in it. Insert the rope through the hole. Tie the rope in a knot large enough so that it doesn't slip back through the drilled hole.

4. Hang the feeder from an overhead support. The gutter can also be nailed to a fence board instead of hanging. Allow the gutter to tip slightly to one side. This will make it easier for all the chickens to reach the grain.

From bottom to top of page: Black soldier fly larvae, mealworms, garlic powder, calcium, grit, DE powder, black oil sunflower seeds.

CHAPTER THREE: Nutritional Supplements for Your Chickens

Many chicken owners feed their flocks a good quality commercial layer feed. Some may choose to mix a feed from all-natural ingredients. Others, choosing true free range, may rely on the natural insects, grasses, and weeds to nourish the flock.

Many of the feed formulations contain some supplements added to the grain mixture. This may be extra calcium, phosphorus, grit, or added vitamins. Protein heavy rations are available that can help feather growth after molting. The same benefits can be gained by feeding free choice supplements of certain foods and minerals. Feeding calcium and grit free choice allows the chicken to ingest exactly what it needs when it needs it. This is beneficial when roosters and chicks are in the flock. Roosters and chicks should not have added calcium. Adding calcium to the grain might cause problems for them. Offering calcium free choice makes it available to the hens, while the roosters and chicks will just ignore it.

What Supplements Are Good for the Flock?

CALCIUM

Calcium is essential for creating the strong eggshells we want to find in the nest boxes. Stress can cause an interruption in calcium absorption, eventually leading to weak eggshells. Illness, age, and breed can also affect the eggshell. Calcium can be supplemented using one of the following easy methods.

Eggshells

Yes, that's right. When you cook eggs, save the shells. Remove any bits of the remaining membrane. Rinse in cold water and lay on the counter to dry. You can use paper towels, a dish towel, or some newspaper for draining. Let the eggshells dry completely.

To crumble the dry eggshells, you can either crunch them up with your hands or place the pieces in a plastic bag and use a rolling pin to pulverize them into small bits.

Or you can use a food processor, mini-chopper, or mortar and pestle.

Purchase Oyster Shell Supplement

An alternative to feeding eggshells is to purchase a bag of calcium supplement from the feed store. This is usually in the form of oyster shell crumbles. There are some supplements that contain calcium carbonate instead of oyster shell. The supplements are not costly and it's a convenient way to offer the extra calcium. You can also alternate between eggshells and oyster shell supplements.

Grit

Grit is essential to your chickens' digestive health. Tiny pieces of dirt and gravel are ingested, then eventually end up in the gizzard of the bird. Food is ingested and stored in the crop along with the grit. The food and grit trickle into the gizzard. The grit takes the place of teeth and grinds the food. The grit particles help digest and soften grains, grasses, and other nutritional bits. Without any grit in the crop, the chicken may end up with an impacted crop and sour digestive tract. Offering small quantities of chick grit while the chicks are in the brooder gives them a good start on proper digestion. As your chickens transition to the chicken run and coop, offer slightly coarser grit. In many cases, if the chicken run is made of dirt and rocks, the chickens will pick up enough grit on their own and no additional supply will be needed. To be safe, offer a free choice container of grit anyway. When the run is grassy or sandy, the chicken may need some grit to encourage good digestion.

Black Oil Sunflower Seeds (B.O.S.S.), Mealworms, and Grubs

Chickens love these delicious little bits of nutrition! Black oil sunflower seeds are not an essential supplement. They are, however, an excellent way to get healthy protein, oil, and vitamins to your flock. In addition, the chickens will follow you anywhere if you are handing out B.O.S.S.

The same can be said for mealworms, and the black soldier fly larvae referred to as "grubs." Mealworms and grubs are high protein snacks for the flock. These are a favorite for most chickens and ducks, so it works as a training treat too. Do you want your chickens to free range in an area? Sprinkle mealworms and watch the chickens obey your wish! Is trying to get your chickens back in the coop frustrating? Shake the bag of treats and sprinkle some on the floor of the coop.

Mealworms and grubs offer a significant protein boost to the flock, which can help chickens during the annual molt. During molting, the chickens lose a significant number of old feathers and then regrow glossy new feathers. The new feathers are ready to keep them warm during the fall and winter. Feathers are largely made of protein and growing new ones requires a diet rich in protein.

Adding supplements of mealworms or grubs helps the flock recover quickly from the annual molt.

Two common ways to add probiotics to your flock are through apple cider vinegar or fermented grains. Both are well suited to the DIY chicken keeper.

Probiotic Apple Cider Vinegar

Making your own unpasteurized apple cider vinegar from apple scraps and peels couldn't be easier and is a great way to provide probiotics for your flock. Adding two tablespoons of your homemade apple cider vinegar to your chicken's water a few times a week will keep their digestive tracts healthier, reduce their susceptibility to intestinal parasites, and strengthen their immune systems.

MATERIALS AND TOOLS
- Apple peels from 12 to 14 apples
- ¼ cup sugar
- Large bowl
- Thin towel
- Strainer
- Glass storage jars

1. Put the scraps in a bowl and add enough water to cover the solids. Stir in about ¼ cup sugar to the large bowl.

2. Place something heavy on top, such as a plate, to submerge the apple scraps. Cover with a lightweight kitchen towel. Set aside for about ten days to two weeks at room temperature. You should see bubbles developing around the edge. If you see a few moldy spots develop, it is okay to remove them without discarding the

rest of the fermenting peels. Try to re-submerge the solids so this doesn't keep happening. If there is a lot of mold present, discard the batch and start over.

3. Next, using a mesh strainer, drain the liquids into a jar and discard the solids. Put a cloth over the top of the jar and apply the ring (but not the flat lid) to the canning jar. Put the starting vinegar in a dark place for six weeks or more.

4. When the vinegar is ready it should smell like a good vinegar. If it smells like wine or alcohol, it is not ready. Return it to the dark cupboard and wait a bit longer. When the vinegar is ready, replace the solid flat lid and store in the pantry. Vinegar does not require refrigeration.

When adding your homemade apple cider vinegar to your chicken's water a few times a week, the ratio commonly used is one tablespoon of apple cider vinegar per gallon of fresh water.

Fermenting Chicken Feed or Grains

Fermenting grain creates lactic acid, which is beneficial to both humans and animals. Evidence and studies show that adding probiotics from fermented feed to your chicken's diet can increase egg weight and shell thickness. It helps chickens naturally resist pathogenic organisms like *E. coli* and *Salmonella*. Also, the chickens will eat less grain and their bodies will use the feed more efficiently, which will please your wallet. Fermenting increases the nutrients available in the feed and creates new vitamins such as B vitamins, folic acid, riboflavin, thiamine, and niacin.

For this process, be sure to use a bucket that has a cover and that will not leach metal or chemicals into the feed during the fermentation process. If you have just a few chickens, you could use a large glass bowl or a quart or half-gallon jar. For larger flocks, you could use a large food-grade storage bucket or a large glass ice tea jar.

Use enough grain to feed your flock a few times in a week. The same liquid can be reused, so the container can be refilled with grain as you dip out the feed needed.

MATERIALS

- Large glass container or other BPA-free bucket with lid
- De-chlorinated water*
- Grains (you can use layer feed ration, or a mixture of whole grains such as whole oats, wheat berries, hulled barley, flaxseed, or rye)
- Whey (optional)

*De-chlorinate water by letting it sit in an open container for a day.

1. Place the dry grain into the container. Do not fill more than ⅔ full. The grain will expand during the fermentation process.
2. Cover with de-chlorinated water. Add enough water to completely cover the grain. Grain not covered by the water could become moldy. Adding a tablespoon or two of whey will help the process get started but is not necessary. *Lactobacillus* bacteria is already present in the air we breathe so it is already in the jar of feed.

3. Cover the container loosely and leave at room temperature.

4. Stir the grain daily, stirring up from the bottom. If water is being absorbed, add more water to cover the grain. Bubbles should be forming and perhaps a thin skim of milky looking film. This is good!
The grain should be ready to use in three to four days. Smell it. It should have a sour smell similar to sourdough. It should not smell rancid or moldy. Mold is bad. If the fermented feed smells like alcohol, you might be able to rescue it by adding two tablespoons of raw apple cider vinegar to the mix and letting the acetic acid eat the alcohol. (Note that raw apple cider vinegar is the kind with the "mother.")

5. Strain the grain, allowing the liquid to drain back into the jar or bowl. The liquid is teeming with probiotics and can be used a few times to make more batches of fermented grain. Add more de-chlorinated water as needed. You can also feed the fermented grains with the liquid and start the next batch from scratch.

How to Offer Supplements to the Flock

When I give my chickens supplements and treats, it's easy to waste much of it. That is the nature of chickens; they scratch and kick as they eat, scattering food and supplements all over the ground where it gets trampled.

Hanging feeders are one way to prevent so much grit, calcium, or treats from being spread all over the ground. Most hanging feeders are intended for the actual chicken feed and will be too large for the amount of supplement served at one time. The solution might be as close as your kitchen cabinet.

Large ladles often have a hole in the handle for hanging. Hang the ladle in the coop, fill the scoop with eggshells, and you have a raised supplement feeder! Even if you don't want to give up the ladle from your kitchen, finding an inexpensive ladle is easy. Check out dollar stores, secondhand stores, and flea markets. Keep an eye open for anything that can be hung on the wall or suspended from the ceiling. Keeping the product up near the chicken's chest level will help with the problem of scratched out waste. Mealworms and grubs are rarely wasted, so go ahead and spread those around with wild abandon!

Supplement Server

Chickens will eat out of any dish you place on the ground. Using free or inexpensive materials, you can work out a way to keep the grit and calcium out of the chicken mud. When sprinkling treats or supplements on the ground isn't a good option, here's a plan for you.

MATERIALS AND TOOLS
- 1 log, approximately 18" in diameter (a tree stump is perfect)
- Router or doorknob drill bit
- Power drill
- Chisel or router tool
- Decorative paint
- Small dish

1. Stand a chicken next to the log to determine the height. I made this about chest high on a medium chicken.

2. Mark the center of the log end. Using the doorknob drill bit, cut the hole for the dish.

3. If you have a router, that will make the job easier. If not, continue to chisel out the center of the log end until you have space to set the bowl down in the log.

4. Fill a small dish with calcium supplement or grit. I used a small custard dish, readily available at many discount stores and flea markets.

5. Paint, if desired. I like spoiling my flock and making things fun. If you want to take it up a level, add some cute painted decorations.

This project can be adapted in many ways. If you want a smaller version, shrink the project down. If you want to make a hanging version, use a slice off the log, cut a hole slightly smaller than the bowl you will use. Suspend the feeder at the chicken's eating level.

Soda Bottle Supplement Feeder

An empty soda bottle can be turned into a supplement feeder with just a few easy modifications.

MATERIALS AND TOOLS
- Plastic soda bottle
- Scissors
- Wire, cable ties, or a hose clamp

1. Rinse the bottle and let air dry.
2. Towards the bottom, cut out a square opening (the opening in the photo is 4"x3").
3. Using wire, cable ties, or a hose clamp, attach the bottle to the wall, fence, or wherever you want it to hang.
4. Fill with calcium supplement or grit.

Roosters and calcium feeding

Roosters do not have a need for extra dietary calcium. The amount in a laying hen commercial ration might not be enough to do any harm to the rooster but it's often a subject of concern. Too much calcium in the diet can cause kidney damage in roosters and young, non-laying pullets. The extra calcium required by laying hens is often four times that of non-laying chickens. The best practice is feeding a low calcium feed and offering free choice calcium as a supplement. Do not mix the calcium supplement into the regular feed. Calcium supplements are self-limiting, meaning the hens will eat only what they need.

CHAPTER FOUR:
Herbs and Forage Planters

Fresh herbs will do more for your flock's health and immunity than most treatments and purchased products. In my experience, offering free choice herbs or bringing herbs to your flock strengthens their immunity and their digestive tracts. The more I make herbs a regular part of the flock's diet, the less illness we experience. Herbs offer nutrients often lacking in commercial food. Fresh or dried herbs can be fed to your chickens. If you don't want to grow herbs or don't grow enough to share with the flock, buy natural or organic dried leaf herbs.

Some herbs benefit the chickens just by being in the coop or run. Mint is a good example of this. My chickens don't really rush to eat the mint. They will eat some, but not a lot. It has a naturally cooling effect if eaten on a hot day, so I will toss some to them. But the real benefit I have seen is more of an aromatherapy effect.

Mint is calming and smells good. Placing it in the coop or by the nest boxes can calm the hens during laying, while freshening the air. In addition, rodents, flies, and other pests do not like the scent. Mint, lavender, thyme, and other aromatic herbs are great natural repellents to use in the coop.

Growing an herb garden for your chickens is one way to provide a ready source of nutrition near the coop or even in the chicken run. But how do you keep the chickens from uprooting all the plants and scratching up the herb garden? My home herbal garden is grown on a tabletop to keep pets and chickens from foraging. I clip herbs regularly to add to the feeding program.

What Are the Best Herbs for Chickens?

No doubt about it, adding herbs to your chicken-care routine results in healthier, happier, more disease-resistant chickens. Adding herbal care along with conventional veterinary medicine and care, where appropriate, results in strong immune systems. Promoting healthy digestive tracts in your birds reduces parasite load and increases immunity and disease resistance. In addition, happy healthy hens lay delicious eggs, regularly, and have a healthy appearance.

Herbs are easy to grow, harvest, and store. Even a small container garden of herbs can provide healthy treats for your flock. Use herbs in the nesting boxes, dust bath area, and sprinkle on the ground as a treat. If an injury occurs, you can make a healing balm using herbs, oils, and beeswax.

Many cleaners on the market emit irritating chemical smells when sprayed. Chickens have sensitive respiratory systems and chemicals can easily irritate their breathing passages. When you use herbs to make a spray cleaner, you solve that problem, plus get the benefits of the aromatherapy while cleaning the coop. Here's a list of some of the most beneficial herbs:

Basil—Supports the immune system, repels pests, assists with mucus membrane health

Bee Balm—Aids digestion and respiratory health

Calendula—Anti-inflammatory, anti-fungal, promotes healing of wounds

Chamomile—Has a calming effect, good for general health.

Cilantro—Antifungal, antioxidant

Comfrey—Promotes healing of bones and soft tissue injuries. High protein content. Can be harmful in large doses. Use in a salve or infused oil.

Dill—Promotes respiratory health, antioxidant, helps control diarrhea

Echinacea—Good for respiratory health after illness has taken hold, kicks the immune system into high gear. As with comfrey, use sparingly.

Fennel—Relaxing, good laying stimulant

Feverfew—Anti-inflammatory

Lavender—Calming, repels pests, antiseptic, odor control

Lemongrass—Fly repellent (add vanilla extract for an extra boost of fly repellent)

Lemon Balm—Repels rodents, antibacterial, calming

Lemon Verbena—Anti-inflammatory, repels insects and pests

Marjoram—Egg laying stimulant, good for reproductive health

Mint—Laying stimulant, controls odor, repels flies and rodents, cooling effect

Nasturtium—Antiseptic, Antibiotic, laying stimulant, natural worm remedy

Oregano—Effective against internal parasites, antiseptic, antibiotic

Parsley—Laying stimulant, high in nutrients

Rosemary—Respiratory health, repels insects

Sage—Effective against internal parasites, good for general health

Thyme—Promotes respiratory health, antiseptic properties, repels most insects

Yarrow—Antibacterial, anti-inflammatory, good for respiratory system, healing, repels insects

WILD HERBS AND WEEDS AND OTHER BENEFICIAL BOTANICALS

Garlic—Excellent in the control of internal parasites, aids overall good health and immunity

Dandelion—Good for immune support, high in calcium, gives eggs a rich yellow egg yolk color

Nettle Leaves (Dried)—High in minerals, calcium, and protein

Chickweed—Natural pain reliever, good source of nutrition, high in vitamins and minerals

Pumpkins—Seeds and fruit are thought to contain natural worming properties

Smartweed—Antioxidant, general health benefits, supports respiratory health, controls yeast growth, antibacterial against some pathogens

Herb Gardens For Chickens

Sage -internal parasites, general health

Oregano -internal parasites,
    antiseptic/antibiotic

Lavender -calming, repels pests,
    antiseptic, odor control

Fennel -relaxing, laying stimulant

Comfrey -promotes healing

Nasturtium -antiseptic/antibiotic

Marjoram -egg laying stimulant
    reproductive health

Parsley -laying stimulant
    high in nutrients

Mint -laying stimulant, odor control,
    repels flies, as a treat

Basil -immune support, repel pests,
    mucus membrane health

Thyme -respiratory health, antiseptic

Dill -respiratory health, antioxidant

Cilantro -antifungal, antioxidant

LemonGrass -fly repellent (add vanilla!)

Chamomile -calming, general health

Lemon Balm -repels rodents, antibacterial

Bee Balm -aids digestion, respiratory health

Other Beneficial Herbs

Yarrow
Chives
Echinacea
Lemon Verbena
Feverfew
Rosemary
Calendula

Wild Herbs and Other Plants

Wild Violet -aids circulation
Garlic -control of internal parasites
Dandelion -immune support and rich colored egg yolk, high in calcium
Nettle Leaves (dried) -high in minerals, calcium, protein
Plantain -wound care, inflammation anti-diarrhea
Chickweed -natural pain reliever, good source of nutrition, high in vitamins and minerals
Pumpkins -seeds and fruit thought contain natural worming properties

Helping Chicken Issues with Herbs

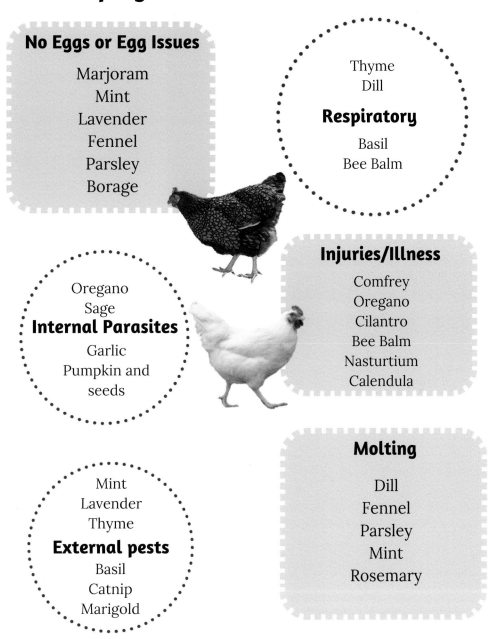

No Eggs or Egg Issues

Marjoram
Mint
Lavender
Fennel
Parsley
Borage

Thyme
Dill
Respiratory
Basil
Bee Balm

Oregano
Sage
Internal Parasites
Garlic
Pumpkin and
seeds

Injuries/Illness

Comfrey
Oregano
Cilantro
Bee Balm
Nasturtium
Calendula

Mint
Lavender
Thyme
External pests
Basil
Catnip
Marigold

Molting

Dill
Fennel
Parsley
Mint
Rosemary

Herbs and Chickens

Adding herb gardening to your chicken care routines will result in healthier, happier chickens. Using herbal care along with conventional veterinary medicine results in strong immune systems, healthy digestive tracts, and a long egg laying life for your hens. Herbs boost immunity, increase resistance to pathogens, and improve overall health and appearance.

Herbs are generally easy to grow, harvest, and store. With even a small garden or container, you can grow herbs to use in your kitchen and have on hand for health problems in the flock.

Where to Use Herbs

Nesting Boxes
Make Healing Balms
Use in Dust Bath Area
Add to Feed
Give as a Treat
Coop-Cleaner Spray

Use For Stress!

Chickens are very sensitive to stressful situations. Attacks by predators, high heat, lack of nutrients, illness, and new flock members can all cause stress and lead to illness. Herbs help combat stress and boost immune systems.

You can quickly construct an herbal snack bar using a shipping pallet. We made two different gardens using pallets. The chickens love pecking off just what they need for a tasty treat that is good for them, too.

How the gizzard works

After the chicken eats, the food travels to the crop. The crop is like a holding tank. It slowly releases food into the proventriculus and then to the gizzard (ventriculus) to be ground up and sent on to be digested further. This is where the grit does its job. The gizzard is a muscular organ that squeezes and contracts, using the rough grit to grind up the food, as teeth would.

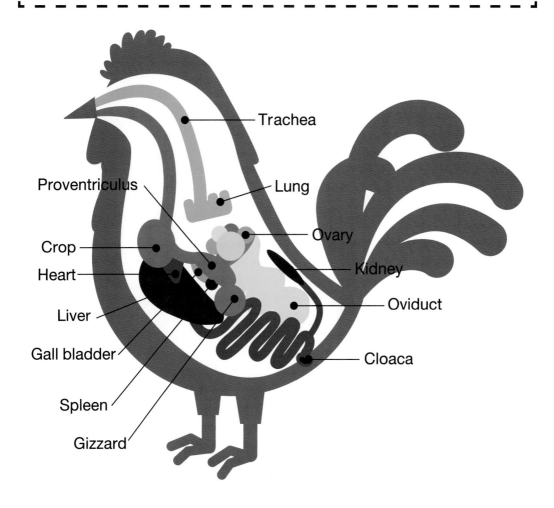

Trachea

Proventriculus

Lung

Ovary

Crop

Kidney

Heart

Oviduct

Liver

Gall bladder

Cloaca

Spleen

Gizzard

Herbal Snack Bar Wall Hanging

For this project, choose a used pallet with the "HT" designation stamped on it. This indicates that it was heat treated and not chemically treated. Some newer pallets are not stamped at all since chemical treatment has been discontinued in most shipping uses.

MATERIALS AND TOOLS

- Pallet
- Black weed-control cloth
- Potting soil
- Wire mesh to cover the garden
- Staple gun
- Herb plants or seeds

1. I chose a pallet that had space to fit my hand into the sections to attach the black weed cloth. Measure the size of the pockets you will need for each section. Double that measurement. I double layered the cloth for extra durability.
2. Fold the cloth section in half.
3. Working from the back of the pallet, attach the cloth, forming a small grow bag or pocket.
4. Repeat the process until you have many pockets made and the pallet is ready to be filled with potting soil.
5. Continue to make the pockets and staple in place from the back of the pallet. Flip the pallet to the right side. Fill the pockets with soil.

6. Pack lightly. Begin adding the seeds or plants to each pocket. If you are using plants, choose smaller sized plants to start the garden. These will fit under the protective wire better and they will grow through the wire quickly.

7. Cut a piece of ¼" welded wire mesh to the measurements of the pallet plus 4" on each end for attaching to the pallet.

8. Do not apply the wire tight up against the pallet. Let it bow outward somewhat. This will keep the chickens from pulling the plants out by the roots. In our example, the wire sits an inch or two away from the pallet. The wire is slightly bowed out from the plants.

9. On the sides, bend the wire covering down and firmly attach to the pallet. You can use larger staples or nails bent over to hold the wire in place.

10. Repeat for the other side.

11. If you planted seeds, let the pallet garden lie somewhere safe from the chickens until the plants are sturdy and growing well. This will give the seeds time to germinate without washing away. Water carefully. Herbs like water but they do not like to be soaked.

We leaned our pallet against the wall, at an angle to keep it from falling. You could also attach the pallet to the fence using screws or bolts and washers.

Small Pallet Border Gardens for Chicken Runs and Yards

After building the larger herb pallet garden, I noticed some sections of pallets left over from other projects. Some of these looked like they would make smaller growing containers for the tiny coops. Even though our chickens get some free-range time each day to wander around and eat weeds, they spend a lot of time in their runs. Placing some growing gardens in the runs helps them choose green plants as a snack. The problem again is keeping the plants growing and the chickens from scratching everything out of the garden.

Much the same as the herb garden pallet wall, I planted lettuce and grass seed in the smaller garden. I didn't put a lot of thought into the type of lettuce. Arugula seeds were handy and they proved to be a quick growing lettuce. The grass seed was a type sold for patching bare spots in lawns. Nothing fancy here.

I let the garden grow for a couple of weeks until the plants were way above the height of the protective wire. In this case, I placed the garden right on the ground.

At first the chickens weren't sure if this was a good thing or a scary thing.

They soon came around and now enjoy the growing garden in the chicken run.

MATERIALS AND TOOLS
- Pallet or section of a pallet
- Potting soil
- Weed control cloth
- Welded wire
- Staple gun
- Seeds (lettuce, herbs, grass)

1. Use the black weed cloth to form pockets and attach them to the back side of the pallet. Attach pockets of the weed cloth to the underside or back of the pallet, using the staple gun to secure. Fill the pockets with potting soil.
2. Plant grass seed, lettuce, or herbs in the pockets.

3. Water carefully so the seeds don't wash away.

4. Measure a piece of ¼" welded wire wide enough to cover the pallet plus extra length to allow it to bow out slightly. This will prevent the chickens from eating the entire plant. They will be able to peck at the part of the plant that grows through the wire covering.

5. Water regularly and hold off putting it in the chicken run until the growth is sturdy and strong.

The first one I made started growing and looked good, so I put it in the chicken run. Well, the plants were not strong enough to withstand the chickens pecking and they pulled the plants out by the roots. The next time, I allowed more time for mature roots to grow before adding the garden to the run.

Plants That Are Toxic to Chickens

Some common garden plants that can cause toxicity in chickens are azaleas, buttercup, clematis, foxglove, henbane, honeysuckle, irises, lantana, lily of the valley, oleander, rhododendron, sweet pea, and vetch. Eating the fruit of the pepper plant is fine for chickens, and they love spicy peppers and the seeds, but the leaves should not be offered or available to your flock. All varieties of potato vines are toxic. Small amounts of cooked potatoes do no harm.

Chapter Five: Chicken Run

The fenced in area connected to or surrounding the coop is called a chicken run. Whether you choose costly fencing or use recycled materials, the goals of the fencing is to:

1. Keep the chickens in
2. Keep the predators out

Keeping chickens in is much easier than keeping predators out. Chicken wire, poultry netting, and picket fences will keep most chickens where you want them. None of these will keep predators from getting to your chickens.

Set up Predator-Resistant Fencing

One common story I have heard too many times from new chicken owners is that they lost some or all their flock to an unknown predator because they didn't have the proper fencing. It is hard to believe that chicken wire can literally be chewed through or torn by persistent raccoons or a fox that wants a chicken dinner, but it's true. Not to mention what larger predatory animals can do to chicken wire. It just isn't strong enough to meet the challenge of a determined carnivore.

There are products on the market that will give your chickens a much safer chicken run. Adding covering over top adds to the protection and prohibits hawks, owls, and other flying predators from taking your chickens.

There are some instances where chicken wire is the perfect choice of wire, but when talking about securing your feathered friends in their run or coop, I do not recommend chicken wire. While it may keep a small flock of chickens in a set area, it is not very strong. Predators can easily move it out of their way, rip it, or tear it open to gain access to your chickens.

Here are a few instances where chicken wire may be used successfully:

- To keep pullets separated from the older chickens inside the chicken run
- To keep chickens out of the garden
- To temporarily plug holes at the fence base line to keep chickens in the run. (Fold or crumple up a piece of chicken wire and stuff it into the hole. Cover with dirt and pack down. Make a more permanent fence repair as soon as possible.)
- To bury underground around the perimeter of the chicken coop and run to deter predators from digging into the coop. Most predators will only try to dig into an area for a short time. When they reach a wire barrier they will often quit digging and move to another spot.

What to Use Instead of Chicken Wire

The preferred wire fencing to use for coop security is called hardware cloth. I am not sure how it got the name because it is much stronger than cloth! It does not bend as easily and is welded, making it a stronger product than chicken wire.

A commonly used method of securely fencing in chickens and fencing out predators is to use hardware cloth attached to a sturdy wooden frame that serves as a side to the chicken run. Make enough frames to surround the area in which you want the chickens to remain. The height of the fencing will depend on if you will cover the chicken run with wire. If you are leaving it uncovered, be aware that the chickens can easily fly over a four- or five-foot tall fence.

At the ground level, make sure the wire extends a few inches underground to prevent predators from digging under the fence. This may seem like overkill if you have not experienced coming home to a large hole dug under the fence and your chickens gone or left injured. We left to go purchase the wire once—it was a short errand to the store and back, so we didn't put the chickens in the coop. We came back to the entire small flock of five gone. A fox had dug into the run and killed our chickens.

As you are attaching the wire to the frames, you can leave several inches excess at the bottom of each panel for extending the wire along the ground, just beneath the surface. The wire will frustrate an animal trying to dig into the chicken run. Most people choose to bring this excess wire out seven to twelve inches along the ground. Backfill the dirt up against the run, completely covering the wire. For more resistance, you can place landscape ties, large rocks, or potted plants along the perimeter.

MATERIALS AND TOOLS
- Sturdy wooden frame
- ½" or ¼" welded wire

1. Bury the wire underground at a right angle from the fence.
2. Backfill with dirt to cover the extended wire.
3. Landscape around the perimeter to provide more resistance to digging and for shade.

Shady Spots for Your Chickens

When planning the chicken run, consider adding plants, structures, or other objects that will provide shade. Shady areas for food and water will protect the chickens from overheating in hot weather and add protection from light rain and snow, too. Add garden plants, large rocks, tree stumps, and other items for perches and activity.

Shade can easily be added with the use of a shade tarp. Shade tarps are dark colored to block the sun's rays. In addition, the shade cloth has tiny holes for airflow. The shade tarp won't keep the rain out, but it minimizes it. I have also used blue tarps and old vinyl tablecloths to add shade in the chicken run.

Large tree limbs that fall during storms are good for shade in the chicken run. The flock will enjoy the new attraction and the insects living in the branches.

Controlling Mud in the Chicken Run

Not immediately, but soon after you move the chickens to the coop and run, any grass or weeds will be eaten and you will be left with bare earth. Unless you have a way to move the chickens around to new grass areas, they will decimate the available greenery. And then you will have mud.

Mud can be controlled by adding wood chips, wood fiber, straw, mulch, stones, and pine needle straw. Swales can be built in the run to control mud runoff.

Drainage issues—When the ground in the chicken area builds up with bedding, dirt, spilled feed, straw, etc., it should be regraded and returned to a somewhat gentle slope towards the downward side of the yard. Natural drainage should be worked with whenever possible. Some folks use a tiller to stir up the dirt and make it drain better.

Runoff—Direct the runoff away from other pens and areas where it can cause more damage.

Grading issues—Often grading issues are to blame for muddy coops. In our coop, the yard has a lot of built up bedding both from mulch and straw and from the coop itself being cleaned out. The chickens love to sift through the leftover bedding but if it's left on the ground for long, it builds up. Regrading is a big job but after a few years of the coop staying in one spot, it may need to be done.

POSSIBLE FIXES FOR A MUDDY CHICKEN RUN

If you find yourself with a muddy chicken run, there are several things you can do to help dry it up.

- Dig trenches and swales to divert the water
- Add stone to the run to help filter out the excess water
- Regrade the area
- Add well-draining fill material to low areas to keep water from accumulating

Here are a few interim solutions to help in a pinch, until you can make a more permanent fix.

Straw—Adding a layer of clean straw to the chicken run cleans off the chickens' feet before they walk back into the coop. Adding a nest of soft straw to the laying boxes will also help keep the eggs cleaner.

Build an elevated boardwalk—We have used pallets with the boards close together, or wide plank boards, as a platform for the chickens to walk on before entering the coop.

Pine chips—Find a tree service that has some fresh pine tree grindings. The ground up trees smell so good and the chickens get a healthy snack, too. Pine needles are a nice treat that helps respiratory tract health.

Wood chips—The square-ish chunks of wood sometimes used on playgrounds can be spread on the ground to temporarily absorb extra moisture. Don't use pine shavings or fine sawdust.

Bales of pine needles—These are more common in some parts of the country than others. This is a great cover for muddy chicken runs. Pine needles hold up well to wet weather. The chickens can nibble on the pine needles, too.

What Not to Use in a Muddy Chicken Run

I have seen pine shavings and sawdust used on top of the run, but this rarely works out well. The shavings just don't stick around and the problem is often worse after these things are added to a muddy chicken run.

Front Porch for the Coop

Building a front porch on the coop can reduce the amount of mud tracked into the coop. The front porch area will collect the muddy footprints and loose debris before the chicken makes it inside and steps in a nest of freshly laid eggs. Keeping plenty of dry bedding, such as straw, right inside the door and near the nest boxes also helps keep dirt controlled.

Pallets are an easy, free (or inexpensive) material to use for a front porch. Look for the pallets that have very little space between the slats. They'll look nicer, and they'll also prevent chicken feet from falling between the slats, causing injury to feet or legs.

MATERIALS AND TOOLS
- Wooden pallet
- Landscape tie or dirt and stone (to level the ground)
- Screws or nails
- Hoe
- Shovel or rake to smooth out the ground

1. You may need to backfill in some dirt to make a level surface, especially if your flock has made craters and hills where none existed before. For your safety and the safety of your flock, build the porch on a level, sturdy surface.

In the example shown, we needed to add some dirt. The ground was very sloped in front of the coop door. While holding the pallet in place at the door, a measurement was taken to see how much buildup was needed. We simply brought in more dirt and wood chips. In front of another coop we used landscape timber to level the surface.

One thing you don't want to have occur is the porch to move around when you step on it. As you can see in the photo, the pallet still needs to be leveled and made sturdy.

After the porch was in place and secured, we backfilled in dirt and stone to keep the pallet sturdy and steady.

2. Nail or screw the pallet into the coop, along the lower door frame. Screws will hold better than nails. In either case, make sure that the nail or screw is completely in the wood and no sharp parts are sticking up where a chicken foot can step on it.

Natural Features for Shelter in the Run

While this fun tree stump hideaway is not intended to be a secure chicken coop, it is a fun place to hang out during the day. If you have a naturally occurring feature in your yard, think about turning it into something the chickens can use for shelter, shade, or play. Shore up the inside with some scrap lumber, add a perch or two and maybe a roof like the one shown here. The umbrella next to it adds more shade and shelter from the weather.

Photo by Ann Accetta-Scott, www.afarmgirlinthemaking.com

Raised Feeding Platform for Food or Water

Building a feeding station covered with welded wire helps keep the chicken run cleaner. Raising the food bowl off the ground can deter the scratching out of feed. The water will stay cleaner if slightly elevated. This is a simple project. You can make an elevated feeding platform to fit the size needed for your chicken run. A small-scale raised platform would make a great addition to a brooder. Chicks are notorious for kicking shavings into the food and water.

CHAPTER SIX: Activities and Boredom Busters

Chickens love to have fun and enjoy new experiences. Adding boredom busters to the run and coop help chickens stay healthy.

Swings

Swings are an easy addition to a chicken yard. Your chickens may be wary of this moving perch at first. Give them time to get used to it. A brave chicken is bound to hop on one day and enjoy the movement. The swings we put together are simply sturdy tree limbs or branches and baling twine from the barn.

Another idea we had was to fashion a log-slice swing. Drill three holes in the wood for the strings to go through. Tie a large knot on the underside to keep the string from pulling though. Adjust the three strings to make them even and securely hang from an overhead support or tree branch. We made one that doubles as a supplement feeder. The hole drilled in the middle holds a small dish of treats, grit, or oyster shell.

Perches

Perches are simply resting places for your chickens. Chickens like being up off the ground while resting. Having plenty of perch options keeps everyone happy. We added quite a few logs and lumber perches throughout our chicken runs. The following perch set up was fun to build from cinder blocks and lumber.

Chicken Jungle Gym

One of our chicken coop runs was boring. The chickens had a large space but basically nothing to do or perch upon. So we made a simple perching set that took little time to pull together.

A.

MATERIALS AND TOOLS
- Cinder blocks
- 2x4s
- Saw
- Paint (optional)

B.

1. Set up the base blocks for the perches.
2. Measure and cut the 2x4 pieces of lumber to the correct sizes for your design.
3. Paint the perch boards, if desired.

C.

Assemble the boards on the cinder blocks. Make sure you have plenty of chicken helpers.

Add a tasty, attractive treat to the area to encourage the flock to test out the new area.

D.

Cabbage Piñata!

A cabbage piñata will keep your chickens busy for hours! This might be something you have hanging around the garden shed already. A metal hanging basket frame makes a perfect spot to place a cabbage, or stuff the basket with other salad greens and veggies. Take out any planting moss that might come with the basket. Hang the basket so it is about beak height. Stand back because a crowd of chickens is going to form!

Pallet Snack Bar

Pallet feeders and gardens are simple to make and, if covered in wire, will continue to grow throughout the summer. As described in the Chapter Four, plant herbs, salad greens, grass, or edible flowers make a healthy treat for your chickens. Directions for making a snack bar from a pallet garden is shown on page 70. Using a similar method, build a slightly raised herb garden or salad garden in the chicken run. Protect the plants from total ruin by covering the garden with ¼" hardware cloth. This will protect the plants and roots from being pulled up completely.

Dust Bath Areas

Dust bathing is how a chicken cleans its body and feathers. The process is essential to the good health of the bird. Have you observed chickens taking a dust bath? The first time you see such activity, you may think the chicken is sick or has been attacked! They lie on their sides, flipping dirt up onto their feathers, all while their legs are digging into the soil to loosen more dirt. It's quite a scene! And apparently the chickens like to do this in groups. So, you may see quite a few of your flock enjoying bath time while a vigilant rooster stands guard.

The dust from all the dirt repels and eliminates lice, mites, and other skin irritants. All that dirt is good for the chickens. Even if you don't provide a dedicated space, the chickens will find a way to take a dust bath. From a young age, chicks will happily roll in a bowl of dirt and begin kicking the dirt up onto their backs.

Making up a dusting mixture for your chickens is simple. These are the components:

Dry Dirt—The main component in the dust bath is regular earth. Find a spot where the chickens have been taking a bath and feel the consistency of the soil. That is what you are aiming to create. I actually use buckets of the soil from an area that the chickens have chosen. After all, I already know that they love it.

Wood Ash (from a fire pit or fireplace)—I add a small bucket, 1 gallon approximately, to the large dust bath. Make sure the ash is from wood and not treated charcoal briquettes. Wood ash contains vitamin K and it absorbs toxins. It's okay if you see the chickens nibbling on a bit of charred wood. The charcoal will help their digestive tract expel parasites and toxins.

Diatomaceous Earth (food grade)—For a large bath, add 4 cups of food-grade Diatomaceous Earth (DE) powder and mix it in thoroughly. The compound acts by dehydrating the insects due to the silica in the DE. This is an extremely lightweight powder; it should not be used in a dust bath without the other ingredients. While the DE powder is all natural and considered harmless, breathing too much of it can lead to respiratory irritation. When adding DE powder to the mix, wear a mask and keep the chickens away from the area until the dust settles. Once the dirt and sand mix with the wood ash and DE, it becomes less powdery.

Builder's Sand—If your dirt is very sandy, you may not need to add more sand. The important factors are coming up with a light, fluffy soil but not so light that it will harm the chicken's respiratory tract.

ADDING HERBS TO THE DUST BATH AREA

Many natural chicken keepers are advocating the use of herbs in almost every area of chicken care. The dust bath is a perfect opportunity to incorporate herbs in your flock's life. The herbs also repel insects and offer tasty bits of nutrition for the chicken as it grooms. The aroma from the herbs will calm the chickens. We all want calmer, healthier, more productive birds in our flocks.

CONTAINERS FOR THE DUST BATH

Small children's pools make a great dust bath if you have the space in the run for one. An old shallow water trough, metal washbasins, or an old dresser drawer can be repurposed into a dust bath, too. If you have access to some logs, cut them into short pieces and stand them on end, in a circular pattern. Add the dust bathing mixture to the circle. The chickens gain both perching places and a dust bath with this setup.

Here are some suitable dust bath containers:

- Large truck tires
- Large wooden drawer or crate
- Old metal laundry basin or wash tub
- Old bathtub
- Child's wading pool
- Leaky water trough or feed trough. The shallow ones work best for a dust bath container.

Try This!

If you want the container to sit lower, dig a hole the size of the dust bath and partially submerge the dust bath in the ground. It will still have sides standing above ground level but will be lower to the ground.

COVERING YOUR DUST BATH

To keep the dust bath dry, make a waterproof cover. All you'll need is an old shower curtain! (A shower curtain is perfect for a dust *bath*, don't you think?) A tarp thrown over the area will do just fine too, unless you expect strong winds. We do cover the dust bath at night when the chickens go into the coop. A large piece of thin plywood also makes an easy cover for a dust bath.

Constructing a framed cover for the dust area gives the cover more stability. Use four pieces of lightweight lumber such as 2"x3" framing lumber, or lighter.

Measure the size, including overhang amount, that you need to cover. Nail or screw the four pieces together at the corners. Screws will be sturdier than nails.

Attach the tarp, shower curtain, or large piece of clear plastic to the frame using a staple gun or finishing nails; make sure to pull the plastic taut before securing it in place.

Some have suggested drilling holes in the bottom of the dust bath to help it drain water. I have found that the dirt of the chicken dust bath gets soggy and muddy very quickly and won't dry out unless there's plenty of sun and dry weather. Therefore, I think covering the dust bath is the most reliable method of keeping the dirt loose and "dusty."

Shade

Shade is important to the comfort of the flock during hot weather. If you have a choice about where to position your coop, choose shade over full sun. Chickens are not very heat-tolerant—they can withstand cold temperatures better than heat waves. Here are a few ideas for adding shade to the chicken run.

Shade tarp—Make your own shade tarp or purchase one. Often these are sold under the term "kennel covers." You can purchase enough coverage for the entire coop or enough to cover a large corner area of the run. Place at least one water source in the shade and make sure the chickens have fresh, cool water during extra hot weather.

Patio umbrellas can be repurposed for shade in the chicken run. Stand an umbrella in a hole, or use an umbrella stand sold at home improvement stores. Another idea is to grab a large flower pot and a lot of rocks. Stand the umbrella in the pot. While someone holds the umbrella steady, add the rocks to the flower pot to stabilize the umbrella.

Large tree branches that fall during a storm can offer temporary shelter and shade. Make sure the tree is not a species toxic to chickens. A large poplar tree limb with branches and leaves leaned against the fence will offer cooling shade and a snack as the chickens nibble on the leaves.

Wooden lattice—The lattice typically found on decks and patios can offer some shade in the chicken run. Secure it to the top posts of the run's fencing, being sure it is securely fastened to the fence so it won't come down on the chickens.

Plant non-toxic garden plants around the outside of the chicken run. The flock will snack on leaves and blossoms that poke through the fence into the run. Consider the following shrubs and plants:

- Herbs such as mint can be trained to grow up the fence, providing shade and treats, while repelling insects and rodents.
- Rose bushes
- Sunflowers
- Marigolds

Security and Fencing

In addition to comfort and activities, the goal of the chicken run is to protect the chickens. Free ranging is great until you start losing chickens to predators. Having a chicken run is helpful for our flock and our family. When we can supervise, the chickens come out of the run and hunt for bugs, play in the compost pile, and feast on delicious weeds. The rest of the time they are confined to a large fenced area surrounding their coop. The run contains what they need and offers safety and security.

Choosing the right kind of fencing is key. While chicken wire may sound like the perfect fencing material, it's actually a poor choice. Chicken wire will keep your chickens inside the run but it won't keep predators from getting inside. It's easily ripped apart by many animals.

Welded wire is a much stronger choice. Welded wire, sometimes called rat wire or hardware cloth, comes in a few sizes. The welded wire with ¼" squares is the best option as it's plenty strong and will also keep rodents from getting in. However, it's also the most expensive. Half-inch wire or even 1" wire will be fine for the chicken run fence.

You can construct the run out of a simple wooden frame and welded wire. Bringing the wire down and continuing it along the ground for another 8 inches adds more protection by preventing animals from digging under the fence.

Chain-link fencing is a strong option, but you will need to add buried wire at the lower third of the fence, around the perimeter. Predators can dig under fencing and gain entry to the chicken run. The photo shows how we attached chicken wire at the lower part of the chain-link fence and extended it down into the dirt. Bend the wire outwards along the ground or under the surface a  few inches. Backfill with dirt to cover the wire. Most predators will dig for a few minutes and then try a different spot. If they can't dig under the wire, they are more likely to leave.

Backfill with dirt to cover the wire. Predators attempting to dig under the fence will be frustrated by the wire and hopefully leave your chickens alone.

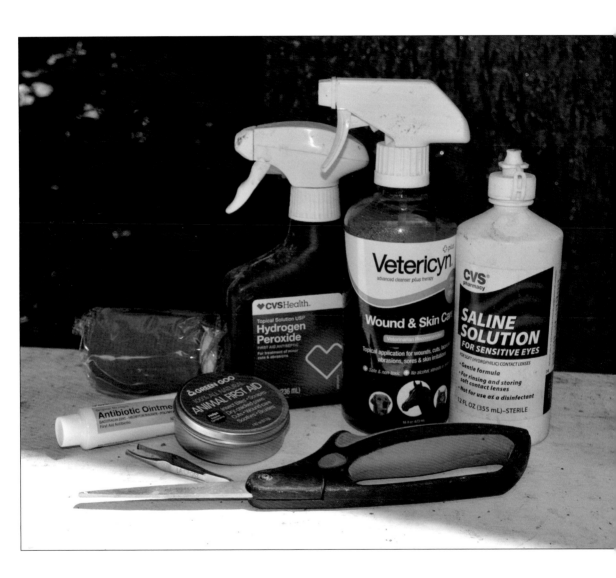

CHAPTER SEVEN: First Aid

Nothing in the following section is meant to replace qualified veterinary advice. Rendering first aid to an injured or ill chicken is intended to ease the chicken's suffering and to stop further damage or symptoms. If the chicken continues to show symptoms, consult a veterinarian promptly. Often, we are raising our herds and flocks in more remote areas without easy access to a veterinarian. Being prepared with first aid knowledge and basic supplies can literally be the difference between life and loss.

It's an awful feeling when you find one of your chickens hurt or ill and you don't have anything on hand to begin treatment. Many first aid kits for livestock or general purpose first aid kits will have some of the items necessary to provide emergency chicken health care. However, it's better to build a kit that will specifically meet the needs of your chickens.

Signs That a Chicken Is Unwell

First, it is important to recognize normal behavior for your flock members. Chickens are almost always busy going about their daily business of eating, scratching, and searching for more to eat and more to scratch. They interrupt this activity to take a dust bath or bask for a moment in the sunshine. Each member of the flock will have its own personality that shines through in these activities and it pays to take note of this. That way, when you notice a marked difference in behavior or lack of activity, you can be alert to a possible health problem or injury.

Initial signs of illness can include droopy appearance and standing off from the flock, lack of appetite, absence of egg production, loose runny poop, swelling of the crop or other body part, or discolored/pale comb or wattles.

Early Care Is Essential!

Chickens will hide the signs of illness so that they don't look like easy prey. It also protects them from being picked on by flock members. Because they hide the signs of a problem, chickens may already be in grave condition by the time you notice anything, so you'll need to act quickly. It pays to be prepared with a first aid box or kit that provides quick relief for the chicken. You will lose precious time if you need to run to a store or wait for the veterinarian to return a phone call. For any life-threatening emergency though, I would still recommend placing that call to the vet. Make the call and then administer first aid while you await further instruction.

Chicken First Aid Kit

Many of the products I have listed here will have a very long shelf life and will not often require replacing. However, it's a good idea to occasionally review the contents of your farm first aid kit and replace stale or old products. Being prepared on the farm or homestead will help you to feel more secure and self-reliant as you go about your day-to-day routine.

Here's what I keep in my farm first aid kit:

Saline Solution—Can be used to clean debris or a speck out of an animal's eye. I use saline solution frequently to irrigate a cut or a sore so I can get a better look at what is going on. Saline solution is inexpensive, available almost everywhere, and not harmful to animals or humans.

Scissors—A knife can be used for most cutting jobs, too, but for cutting bandages, I prefer a pair of scissors.

Disposable Vinyl Exam Gloves—Even if you are not squeamish, some things are better off not being handled with bare hands. Infections in wounds are one thing that should be treated while wearing disposable gloves. Also, wearing gloves protects the animal from further infections being introduced to it while it is already fighting something.

Bandages, Gauze Pads, Band-Aids, Tapes, Vet Wrap or Co-flex Bandage—Buy an assortment of sizes and types of bandages. One bandage might not work on all types of wounds. Vet wrap is perfect for wrapping over the bandage because it is sticky and helps the bandage stay in place. When the weather is wet, I add a piece of electric tape to help keep everything in place even longer.

Wound Care Sprays and Ointments— No farm first aid kit is complete without Vetericyn, Neosporin, and Blu-Kote. These are products I use frequently. Blue-Kote is an antibacterial and anti-fungal spray that also coats the area in blue color. Particularly in chickens, a red wound can be an open invitation to pecking and cannibalism. Spray even a small red area with Blu-Kote to avoid the issue. If the wound is close to the eye or mucus membranes, spray a cotton swab and then dab it on the area. Vetericyn is a commercially available wound spray. In my opinion, it is worth the rather high cost. We keep it on hand and use it for all the animals and pets on our farm. Neosporin is widely available. Be sure that you buy the original formula and not the type with pain reliever in it. The pain reliever is often toxic or not tolerated well by chickens and other animals.

Scalpel—You may not want to use a scalpel but it is good to have a few new ones in the farm first aid kit. Surgical cutting on your livestock is best left to veterinarians, but I keep a scalpel on hand in case I know the vet won't get there in time.

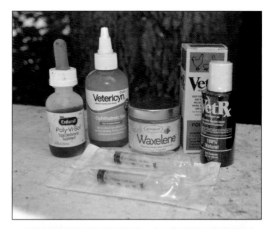

Store your first aid supplies in a waterproof container with a secure lid. Storage tote boxes are a good choice for this, along with a bucket and tight-fitting lid. Use a container that is easy for you to grab and take to the injured or sick bird.

Other Useful Things—Hydrogen peroxide, iodine/Betadine solution, tweezers, cotton swabs/Q-tips, and a syringe (for giving oral meds or even yogurt) will round out your kit.

I can fit just about anything I need to tend to an emergency in this small paint bucket with lid.

Household Products for Common Ailments

Let's face it—we could easily spend a lot of money on retail products aimed at barn-yard animal care. Some of these dollars would be well spent. There are also many common kitchen products for the barn and coop that can do the job just as well as commercial products, and without chemicals or unwanted additives—or the daunting price tags.

IMPACTED CROP

Olive Oil—Chickens can benefit from olive oil if they have an impacted crop. Using a small syringe, carefully open the chicken's beak (or if you are lucky, she will open it for you). Slowly push the syringe so the oil drips into the mouth. Don't squirt forcefully because you could force some into her lungs by accident. Massage the crop after the hen swallows the oil. This will help break up the clump so the crop can pass the material through to the gizzard.

EGG-BOUND HENS

Olive Oil—Another use for olive oil happens at the other end of the chicken. Occasionally a hen will strain to pass an egg. It might be an extra-large egg or she could be older and not as elastic as she once was. Coating the vent with a thin smear of olive oil can assist her in passing the egg.

Olive oil is a good way to add calories in an animal that has been undernourished. Do not over do this, though! Fats should still be the smaller nutritional component of the diet. For example, you could safely mix 1 teaspoon of olive oil into ½ cup of layer feed.

Epsom Salt—Epsom salt is a good source of magnesium. Soaking in a bath of warm epsom salt and water relaxes the muscles of an egg-bound hen. It is also good for soaking bruised legs or feet. Chickens with bumblefoot abscesses can have a foot soak in Epsom salt and water as well.

Blood Stop Products

Few events are quite as scary as blood rushing from an animal's wound. Animals hurt themselves, each other, and sometimes they hurt you. It's a part of barn life that shouldn't happen often, but when it does, you will want to be prepared. Keeping some or all of the following products in an airtight bin in the barn storage room might save a life one day.

All of the following can act as blood-stopping treatments. Once the blood flow is staunched, you can treat and bandage the wound. These products would be applied directly to the wound. Cover with a compress and apply light pressure to hold in place.

- Cornstarch
- Vinegar (it works but it might sting a little)
- Turmeric
- Tea bags (moistened)
- Sugar
- Yarrow herb (crushed or chopped fine and placed on the wound)

Infections/Wounds/Internal Parasites

Oregano offers natural pest-repelling and natural antibiotic action. Oregano can be fed fresh or dried. Add to food, sprinkle on the ground, in nest boxes, or mix into homemade treats.

Garlic aids the gut in staying healthy and repelling parasites. Use it fresh or dried in small quantities throughout the year.

Honey can be used as an antibacterial and healing ointment. Use raw, unpasteurized honey for the healing benefits.

Coconut Oil is one of the best treatments for skin irritations. Coconut oil has healing properties and coats and protects abrasions.

Molasses adds calories and some nutrition for a weakened chicken.

Herbal Salve for Healing Chicken Wounds

Some people prefer to stay away from commercial pharmaceutical products due to preference or sensitivities to certain ingredients. Making a simple wound care salve is quick and easy. Keeping a few ingredients on hand allows you to make various herbal salves that specifically help solve the problem you are treating. So many of the herbs we grow in our gardens have healing medicinal properties. Skin wounds, inflammation, and even infections can be healed using natural herbal salves you make in your kitchen. This recipe makes enough for two 2-ounce tins or pour into one 4-ounce jelly jar.

MATERIALS AND TOOLS

- Glass jar
- Healing herbs, dried (Suggested: chamomile, calendula, dandelion, lavender, plantain, thyme, wild violet)
- Olive oil, sweet almond oil, or grapeseed oil
- Mesh strainer
- ½ oz beeswax
- ½ oz coconut oil
- Tea tree essential oil
- Oregano essential oil
- Vitamin E capsule

1. First, prepare an infusion using healing herbs and oil. Fill a jar about ⅔ with the dried herb. You can make combination infused oils but I prefer to make each infusion separately. Always use dried herbs and botanicals when making infused oils. Pour oil of choice over the top of the herbs.

2. Place the jar in the sun for a few days. If you don't want to wait that long, place the jar in the top piece of a double boiler. Bring the water to simmer, and then turn off the heat. Let the herbs or flowers in the jar of oil sit and warm for a few hours. Or, set the jar in a few inches of water in a slow cooker. Let the oil warm for a few hours.

3. Strain the oil. I used a fine, mesh strainer. Add a piece of cheesecloth if you think it is necessary. Store the oil in the refrigerator until you are ready to use it in a salve or lotion recipe.

4. To make the salve, melt the beeswax and coconut oil together in a double boiler or glass canning jar placed in simmering water. Add 4 oz of the infused oil (I recommend 2 oz of calendula oil and 2 oz of another healing herbal oil such as lavender or dandelion).*

5. When these ingredients are completely melted, add 12 drops of tea tree oil, 12 drops of oregano essential

oil, and the contents of the vitamin E capsule. Vitamin E acts as a preservative for the salve. Stir to combine and then quickly pour the mixture into the small glass jars or small metal containers.

6. The salve will harden quickly. Since it has so much coconut and olive oil in it, when you take a scoop to use the salve it melts quickly and absorbs quickly into the skin. After the oil has penetrated the skin, the salve continues to work and heal.

   Use the salve for cuts, scrapes, pecking wounds, predator bite wounds, and minor infections. No homemade recipe takes the place of proper veterinary advice. When you are treating a wound that is not improving, seek the advice of a qualified veterinarian.

*If the injury is a sprain or broken bone, use 4 oz of comfrey infused oil instead of another infused oil. Comfrey has properties that support healing of bone and soft tissue injuries.

Drawing Salve

Abscesses, deep infections without broken skin, and cysts may benefit from a drawing salve such as this one. I also use drawing salve to treat a bumblefoot infection. Occasionally the cyst that forms can be deep and take a long time to actually form the kernel of infection. Drawing salve can help draw the infection to the skin.

Apply the drawing salve ointment to the chicken's foot. Cover with a 2"x2" gauze pad. Wrap with cohesive bandage (often sold as vet wrap). Change bandage and ointment daily.

Splinters are another good use for a drawing salve. Use on the chickens' feet or on your hands!

INGREDIENTS
- 6 tablespoons olive oil infused with plantain, calendula, or goldenseal (see page 100 for instructions on making an infused oil)
- 3 teaspoons beeswax pellets
- 3 teaspoons activated charcoal
- 3 teaspoons bentonite clay

- 30 drops lavender essential oil
- 15 drops tea tree essential oil

1. Combine infused oil and beeswax in a canning jar. Set the canning jar into a few inches of water in a slow cooker or place in a pan of water on the stove. This creates a makeshift double boiler.
2. Heat until the wax is melted. Remove from the heat.
3. Stir in the remaining ingredients, clay, charcoal, and essential oils. Pour into a 4-oz canning jar or a 4-oz tin.
4. Let cool. Cover with lid and store in a cool, dry location when not in use.

Hen Saddles

"Why is your hen wearing a dress?" I get this question a lot. It's not really a dress; the hen saddle (sometimes called hen aprons) protects the chicken's back and feathers from the treading of a rooster. If you don't keep a rooster, you may not ever need the protection of a hen saddle. Making the hen saddle is an easy DIY project. But first, let's talk about why the hen saddle is good for the hen.

Observing chickens mating can be disturbing if you haven't seen it before. Roosters are not gentle when they mate. The hen submits by crouching down. The rooster jumps on her back and treads his feet into her feathers to gain his balance. The actual mating is quick and then both hen and rooster shake their feathers, walking off to continue foraging. The rooster may go from one hen to the next in quick succession. And if you have more than one rooster, the boys may have their own idea of which hen belongs to each of them. Roosters have an interesting idea of courtship!

Rooster feet are large and the talons are sharp. In addition, the spur, which is a defense feature growing out of the rooster's lower leg, may be quite long. All of these structures are digging into the back of the hen while the rooster is mating.

Feathers are meant to protect and fluff, but they cannot always withstand repeated abrasions. The mating behavior can cause the hen to lose her back feathers. After the feathers fall out, the hen is still a willing victim in the mating game. Now however, the skin on her back will take the wear and tear—plus, the tender, exposed skin may get sunburned.

Some hens seem to have a lighter feathering and lose their feathers quickly. Some manage to keep a downy covering.

The first sign of feather loss starts near the tail of the hen; you'll notice that the tail looks downy instead of fully feathered. Assuming it's not molting season (late summer or early fall), feather loss is probably a sign your hen could use some extra protection.

Rooster-caused feather loss is usually seen in the spring. Mating season begins as the days lengthen. Look for feather loss at this time and think about using a saddle to protect the hen. As fall approaches and the chickens begin the yearly molt and feather regrowth, remove the hen saddles. The saddles, if left on the hen, can abrade and loosen the new growth. Mating behavior should have calmed down by this point in the year.

Using a hen saddle will protect the feathers before they fall out. If you don't want to sew a hen

saddle, there are many options for buying them. But if you can follow a simple sewing pattern, you may enjoy stitching up a few saddles to protect your hens.

Depending on the temperament of the hen, she may object to being caught and held while you dress her. After the saddle is on correctly, the wings fold over most of it and the hens hardly seem to notice they're wearing one. Occasionally the saddle will roll up the hen's back. Flip it back down and when she adjusts, her wings it will cover it up again.

DIY Hen Saddle

To make these hen saddles, I used two contrasting fat quarters, which can be purchased wherever quilting supplies are found. Two fat quarters will make four hen saddles. Of course, you can use any leftover cotton fabric you have on hand, too.

MATERIALS AND TOOLS
- 2 pieces of cotton fabric (8½"x9½" each)
- 12" length of ½" wide elastic
- Scissors
- Sewing machine or needle and thread

1. Cut two pieces of cotton fabric in the shape of the pattern.
2. Place the two pieces of fabric together, right sides facing each other.
3. Stitch the two layers together, leaving the opening shown, unsewn for turning. Use a quarter inch seam allowance and clip the curves to make turning easier.
4. Turn the hen saddle to the right side by pulling the saddle through the opening. Iron the saddle to make it smooth.

5. Turn the opening raw edges to the inside. Iron again. Fold over the top for the elastic casing. Sew to the body portion at the top, making a casing for the elastic band.

6. Insert the elastic band through the casing. Attach to each side of the hen saddle and stitch in place. If you are having trouble threading the elastic, attach a safety pin to one end to push through the casing.

B.

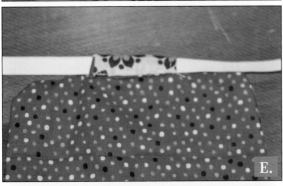

C.

D.

E.

F.

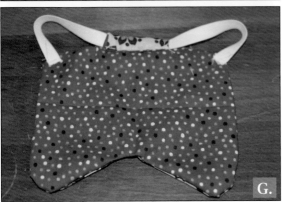

G.

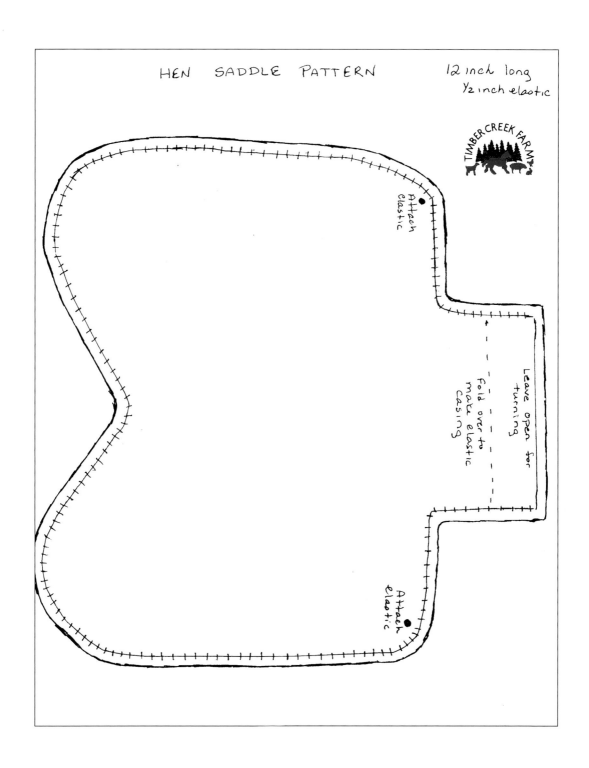

HEN SADDLE PATTERN

12 inch long
½ inch elastic

TIMBER CREEK FARM

Attach elastic

Leave open for turning

Fold over to make elastic casing

Attach elastic

CHAPTER EIGHT: Basic Coop Maintenance

Installing a Box Fan

Box fans in the coop increase ventilation, decrease fly activity, and remove odor. Small fans won't do much good in a large shed or large coop. The standard box fan may not have the power to withstand the dust and feathers. Buy an industrial or farm-grade box fan for better results.

You want to hang the fan so that it is not directly blowing on the roosting areas. Chickens should not sit in a draft even in hot weather. In one of our coops, the fan is hung over the door of a large garden shed turned chicken coop.

Make sure you securely hang the fan high enough that the chickens will not fly into it as they leave the coop.

Use chain or sturdy strong rope to hang. We had to tilt ours back and up to clear the doorway.

When plugging into the electrical, made sure the connection is not accessible to the chickens.

Installing a Heat Lamp for Chicks

The question I have the hardest time answering involves the use of traditional heat lamps in the chicken coop. There are a number of reasons why I hesitate to get behind this practice. The most important reason is that it is a fire risk. You are adding an extremely hot electrical item into an environment of flammable bedding. Straw, pine shavings, and most likely a wooden structure will burn quickly. Every year we hear of tragedies started from the use of heat lamps in a coop.

If you absolutely have no safer place to brood the chicks than in the coop, take extreme care when hooking up the heat lamp. Keep it out of reach of the chickens.

Suspend it from a sturdy rafter or chain, hung over the brooder. Keep all straw and shavings away from the lamp. Sweep away any cobwebs from the ceiling.

If you have a safer space to brood the chicks, I recommend doing it there. A storage tote or brooder pen, in your garage or laundry room, might be an answer to the brooding location. The garage most likely has a concrete floor. Check the lamp frequently. If you must use a heat lamp in the chicken coop for brooding chicks, be extremely vigilant.

There are products on the market that are much safer for providing warmth to the chicks. Lights with guards surrounding them, shelf-style chick warmers, and warming pads for the floor are a few of the newer options.

Checking for Proper Ventilation

Ventilation is a term we hear often when discussing the chicken coop. How do you know that your coop is ventilating properly, is draft-free, and is not accumulating dust and moisture?

Dust accumulating, stale air, and ammonia odor are signs that the coop is not properly ventilated. A pre-built coop will most likely have roof line vents installed. It may even have a ridgeline vent underneath the shingles. If you built a coop from scratch, you may need to add more ventilation to the roofline. This can be easily accomplished by cutting a small opening near the roof. Cover the opening with sturdy pieces of hardware cloth. Even a small hole can be an open door to predators and rodents.

Clean the droppings area frequently and make sure no moisture is accumulating in the corners of the coop.

DIY Coop Air Conditioner

An air conditioner for chickens? Why would anyone need to air condition the chicken coop? In some areas of the country and in some coop designs, this may not be a frivolous idea. Understandably, people worry that their chickens will be too cold during the winter months. But in fact, chickens are prone to sudden death from heat illness more than problems with cold weather. A properly constructed coop will protect the birds from most cold weather quite well. The heat is hard to get relief from. When a heat wave hits early in the season, and the chickens have not had time to acclimate to the warmer temperatures, they can easily suffer heat stroke, dehydration, and death.

Building a simple air cooler, even one that runs on batteries, can make a difference in the health of the flock during a heat wave. For areas that have no access to electricity, there are battery-powered fans on the market that use rechargeable battery packs, similar to cordless drill batteries. You may not want to use the AC on a regular basis, but in a serious heat wave, when your chickens are suffering, this can cool the birds down enough to save their lives.

MATERIALS AND TOOLS
- Small electric fan or rechargeable and battery run fan
- 2" or 3" PVC pipe elbow
- Skill saw or sharp knife to cut the holes in the cooler.

1. Cut two holes in the top of a cooler, as shown. If the cooler is made of Styrofoam, use a knife to cut the holes.
2. Place the fan face down on the cooler top and draw the outline of the fan's face.
3. Draw the outline of the PVC elbow pipe at the end of the cooler top.
4. Insert the PVC fitting into the hole.
5. Lay the fan face down on the top of the foam cooler.
6. Fill the cooler with ice. Plug in the fan. As the air from the fan is forced into the cooler, it will force cool air out through the PVC pipe and into the area you want to cool down.

Cleaning Chicken Water Containers

In the heat of summer, nothing is as important to our chickens and livestock as clean drinking water. At least once a week, I give the water containers, buckets, bowls, and founts a good scrub. It doesn't take long and helps maintain a healthy environment. Gathering the supplies takes only seconds. White vinegar, water, and a scrub brush are all I use. The non-toxic coop cleaning spray can be used if you are concerned about additional germs from illness or for any reason.

Sediment from rusty water or algae growth.

First, dump out any remaining drinking water. Add vinegar to the bowl. Spray with Herbal Vinegar Coop Cleaner (page 113) if needed. Let it sit a few minutes. Scrub with a brush to loosen all the gunk, algae, and rust sediment. Rinse well. The bowl or water container should smell much better, look cleaner, and be ready for a refill.

In between scrubbing, make sure you refill with clean water every day. The algae and rust form a biofilm in the water, which affects the taste. Few of us would choose to drink a big glass of stale, smelly water, and our chickens and livestock agree. They may drink some, but not enough to combat the potential dehydration. If you see that the water containers you use have not been depleted much during the day, chances are the water is foul. Dump it out, clean the container, and refill. Hopefully that will make a big difference in the amount of water your animals consume.

Herbal Vinegar Coop Cleaner

The respiratory system of a chicken is highly sensitive. Chickens are prone to respiratory problems from dust, chemicals, and other environmental hazards. A buildup of fumes from chicken droppings, spilled water, and cleaning products irritates the delicate breathing system and leads to illness. Cleaning the coop with a harsh, irritating cleaner can be dangerous for the respiratory tract of your chickens.

Using a coop cleaner made with natural products will not only clean but disinfect and deodorize the coop, without irritating the chicken's respiratory system. With simple, easily found items, you can mix a quart of non-toxic cleaner that will last through many coop cleanings.

INGREDIENTS AND MATERIALS

- A few handfuls of mixed herbs. You can use fresh or dried herbs. Choose from mint, lavender, thyme, marjoram, and oregano.
- Vanilla beans
- Cinnamon stick
- Citrus peel from two or three lemons, oranges, grapefruit, or limes (peels from citrus fruit can

be stored and frozen in plastic bags until you need them)

- White vinegar (do not use apple cider vinegar for this cleaner)
- Quart canning jar and lid
- 16 oz spray bottle
- Cloth for straining the liquid or a fine mesh strainer

1. Place the peels from the citrus fruit in the jar. Add the herbs, vanilla beans, and cinnamon stick to the jar. Completely cover the ingredients with white vinegar. You can substitute inexpensive vodka if you prefer.

2. Put the lid on the jar. Shake gently to distribute contents. Place it in a cupboard or under a dishcloth for two or three weeks. Shake the jar occasionally while it steeps.

3. Strain the solids from the liquid using either the cloth or fine mesh strainer. Pour the liquid into a spray bottle. Save the extra cleaner in a glass jar for when you need a refill.

Managing Chicken Manure and Coop Waste

Composting chicken manure is a side benefit of raising chickens. Chickens provide us with hours of companionship, fresh eggs, and . . . manure! Lots of manure. Approximately one cubic foot of manure is produced by each chicken in approximately six months. Multiply that by the six chickens in an average backyard flock and you have a mountain of manure every year! If you lived on a farm, that

Photo by Laurie Neverman, http://commonsensehome.com

may not be a problem, but in a small backyard or in a neighborhood, there has to be a plan to take care of the chicken manure. How can you turn your pile of chicken manure into something beneficial? With a little extra effort, the manure can become rich compost for your garden—and maybe you will have enough to share with the neighbors, too!

Cautions for Composting Chicken Manure

Most chicken owners know that fresh chicken manure can contain *Salmonella* or *E. coli* bacteria. In addition, the fresh manure contains too much ammonia to use as a fertilizer and the odor makes it unpleasant to be around. But, when properly composted, chicken manure is an excellent soil amendment. Compost does not have the unpleasant odor. Chicken manure compost adds organic matter back into the soil and contributes nitrogen, phosphorus, and potassium to the soil.

Two Reasons to Compost the Chicken Manure

1. Adding the manure directly to the garden can spread pathogenic organisms to the soil, which can be picked up by low growing leafy greens and fruit.
2. Fresh manure will burn the plants roots and leaves because it is too strong. Once composted, it will help your plants grow better.

How to Compost Chicken Manure

The waste you scrape out of the coop, including all the shavings, sawdust, straw, and hay can be added to the compost bin with the fresh manure. Compost components are usually labeled either brown or green. The bedding materials, along with any additional yard plant debris, leaves, small sticks, and paper would be your brown parts. Manure and kitchen scraps would be the green parts. When composting chicken manure, a ratio of two parts brown to one part green is recommended because of the high nitrogen content in the manure. Place all the materials in the compost bin or composter. (The bin should be about 1 cubic yard.)

Turn the Compost Pile

Mix, stir, and turn the composting material regularly. Occasionally check the inner core temperature of the material. A temperature of 130 or up to 150 degrees Fahrenheit is recommended to allow the soil bacteria to break down the pathogenic bacteria from the manure. Turning and stirring the pile allows air to enter to keep the good bacteria working. After approximately one year, you should have some

Photo by Heather Gibson

Photo by Heather Gibson

very rich, valuable compost suitable for your garden. All of the *E. coli* and *Salmonella* should have been destroyed by the heat produced during composting. It is still advisable to carefully wash any produce grown in a compost fed garden.

Some people make a two- or three-box system for compost. As the material is turned and mixed it is moved to the next bin.

A Few Safety Precautions

- Always wear gloves when handling manure.
- Do not add cat, dog, or pig feces into your compost.
- Always wash produce thoroughly before eating. Individuals with compromised health should not eat raw food from a manure-fed garden.

Building a Compost System

As you can see in the example photos, a compost bin is easy to make from leftover boards or pallets.

Photo by Michelle Hedgecock, www.mdh-services.com

MATERIALS AND TOOLS

- Pallets or boards
- Nails or screws
- Posts (4x4 works well)
- Hammer or drill driver

1. Gather the pallets or boards. If using boards, make the three or four panels for the sides.
2. Attach three of the panels at the corners using nails or posts hooked to the panels. This makes a three-sided box, open in the front.
3. The fourth panel is used as needed. If you don't want the chickens and other animals to scratch through the compost, close off the box using the fourth panel.

Build a Three-Bin Compost Center

Photos courtesy of Adventure Acres Iowa, http://adventureacresiowa.com.

MATERIALS AND TOOLS

- 7 pallets
- 18 screws
- Screwdriver or drill driver

1. Begin with 7 pallets in good condition.
2. Working from the center of the three-bin set, stand up the two sides and one back pallet
3. Attach the back pallet as shown, making sure that the side pallets are approximately centered half way on the back pallet.
4. Add the second back pallet to the structure, connecting to the first back pallet and the middle side pallet.

5. Attach the next end side pallet. Repeat steps 3 and 4 for the remaining bin on the other side.

G.

H.

Other Types of Compost Bins

A simple compost system can be constructed using only posts and chicken wire. Measure out the area and mark the post locations.

Using a post hole digger, dig the post holes, insert the posts, and backfill the dirt to secure the posts.

Wrap the wire around each bin, pulling the wire taut as you proceed. Attach the wire with hammer and nails or a staple gun.

A sturdier compost system is shown using treated lumber to construct the bins, and an attached, hinged lid.

The photo below shows the mixture of green and brown material that is commonly composted.

CHAPTER NINE: DIY Chicken Feed for Laying Hens

With so many feed choices available in most feed stores and farm co-ops, is it cheaper to make your own feed blend? If you're careful to adhere to nutrition guidelines and you have the ability to store excess grain safely, you can make your own layer feed, although, I don't think you will save money. If you choose to substitute any ingredient for another, pay close attention to the protein level of the substitution. All grains are not created equal in that department.

There will be some specialty ingredients to purchase in addition to the grain and legumes. Some of these products are:

Brewer's yeast (with garlic, if available) is a good source of niacin for strong bones. There are some products on the market that include garlic powder in the brewer's yeast, or you can simply add garlic powder to the mix on your own. Garlic helps by repelling insects such as ticks, mites, and fleas, while aiding the immune system.

Flaxseed is an excellent source of Omega 3 fats.

Diatomaceous Earth (Food Grade DE Powder) is commonly used externally and around the coop for pest control, but food grade DE powder can also be fed to chickens without any harm. Evidence points to this product helping with internal parasite control.

Kelp powder adds vitamins and minerals that are essential to the chicken's diet.

Probiotic powder helps protect the intestines from bacteria and parasites. Probiotics also provide immune support because healthy intestines absorb nutrients better, contributing to better health.

Nutri-Balancer is a commercially prepared ration balancer containing kelp. When using a nutrition-balancing product like this, you won't need to purchase the kelp powder.

How Much Feed Does My Chicken Need?

The average chicken food intake is approximately ½ cup per day per chicken. For our flock, we feed ¼ cup twice a day, per chicken. This is an estimate, of course. Since the flock eats from community feeding bowls, it's impossible to say exactly how much each bird eats.

My recommended DIY chicken rations contain the following ingredients. These are percentages by weight, not by volume.

- 30% corn
- 30% wheat
- 20% dried peas
- 10% oats
- 8% fish meal
- 2% minerals and vitamins provided by Nutri-Balancer

In addition, smaller amounts of flaxseed and sesame seed can be added for variety. Feed supplemental grit and calcium free choice.

Alternately, you can start with a 50 lb bag of layer ration and then add the ingredients below. Add the following ingredients to the 50 lb bag in the stated amounts.

- 40 oz steel-cut oats (a large canister)
- 2 cups cracked corn
- 4 cups sunflower seeds
- 1½ cups brewer's yeast
- 1 cup garlic powder

- 1½ cups probiotic powder
- 1½ cups DE powder
- 1 cup kelp
- Dried herbs if you have them on hand

Mix into the 50 lb bag of layer feed. Feed the amount of ration you normally feed to the flock.

Sprouted Whole Grains for Chicken Food

Sprouting pans of fodder from whole grains significantly stretches your feed dollars. Sprouted grains have more usable protein and other nutrients. The green growth takes about a week to grow from the grains. My method rotates the process, reusing the empty pans, and continually provides fresh fodder for the flock. If you plan on regularly sprouting fodder, grab a large container and pre-mix your grains. Store, covered, in a cool dry area.

INGREDIENTS AND TOOLS
- Bucket
- Strainer
- Something to set the pans on while the grains sprout (the pans will drip excess water)
- At least two large disposable aluminum pans
- An ice pick, cooking fork, or something to poke holes in the bottom of the pans for drainage. If you are using plastic trays, a drill might be a better tool.
- Whole grains (choose or mix from wheat grain, oats, barley grain, dried peas, and flaxseed)

1. After mixing the whole grains, soak overnight. In order to sprout the correct

amount and not waste any grain, I pour the grain into the aluminum pan first, to a depth of half an inch. This is the recommended amount for sprouting without encouraging mold or spoilage in the lower part of the pan.

C.

2. Pour the grain from the pan into the bucket. Cover the grain completely, with water. Make sure all the grain is submerged and the water is above the level of the grain. The grain will absorb water and bloat, so give an extra 2" of water above the grain.

Using the ice pick, punch holes in the bottom of one aluminum pan. Leave the other pan with no holes.

D.

3. Using a strainer or slotted spoon, transfer the soaked grain back to the aluminum pan with holes. Set this pan into the pan with no holes. This will hold any draining water as the grains sprout.

4. Water the pans twice a day. Check that the pans are draining. You do not want the grains floating in water. Don't allow the pans to become dried out.

E.

I used a spray bottle to water the grain and had to reduce the watering to once a day during a very humid weather period. Watch carefully for any mold growth. Do not feed moldy food of any kind to your flock.

Roots will form in one or two days. The grains will sprout quickly, although it takes almost a week for a fully grown crop of fodder to form. The fodder will take on the appearance of sturdy grass when fully grown.

5. Once the roots and fodder have grown, it is easy to pick up the fodder block from the pan to feed to the chickens. Stand back! They are going to love it.

Benefits of Feeding Fodder

The grains contain nutrients for the seed to sprout and grow. Using the grain in dried form doesn't make all nutrients available to the chicken or other grazing animal. Once the grain sprouts, the nutrients have an increased availability during digestion. Since more nutrition is available from each grain, you will not need to feed as much grain. Getting a fodder system set up is simple. Once you have the system in place, keeping it going only takes a minute each day.

CHAPTER TEN:
Treats for the Flock

Treats for Hot Weather

Heat waves are a serious threat to chickens, especially when they come early in the season, before the flock has had a chance to acclimate to the warmer temperatures. Cooling your birds down with some cool, tasty, treats can make a difference in how they adapt to the extreme heat. Make sure there is plenty of fresh water available at all times. Make some of these chilly refreshments for your flock, too.

Whatever Floats Your Watermelon Boat

Remember those Jell-O molds from family dinners? If you don't have one, the watermelon shell works just as well.

- 4 cups cut up fruit, such as watermelon, berries, small bits of apple, peaches, or other soft fruits
- Kale, torn into small pieces
- Frozen peas
- Black soldier fly larvae or mealworms
- Black oil sunflower seeds
- Jello-O mold pan or watermelon shell

Note: Almost anything safe for chickens to eat can be cut up and mixed into the fruit. Use what you have at the time. Switch out the watermelon for a different melon if that's what you have on hand.

1. Chop the fruit into small bites. Do not discard the fruit juice!
2. In a large bowl, mix the fruit, kale, and peas. Pour the fruit and juice into the Jell-O mold or watermelon shell. Add any extra fruit juice to fill the mold.
3. Sprinkle the black oil sunflower seeds and the mealworms or fly larvae on top.
4. Put the mold into the freezer and allow to freeze solid.
5. When frozen completely, serve immediately to your flock for a welcome cooling treat.

Note: Save any rinds and leftover juice from fruit. Freeze the leftover juice for another batch. The rinds can be stored in freezer bags for another frozen treat. The same can be done for any leftover fruit.

Treats for Cold Weather

Cold weather is the appropriate time for feeding your flock extra energy foods like corn, winter squash, and pumpkin. Corn is easy to obtain and mix into any combination of foods that are safe for chickens to eat. Pumpkin is high in vitamins and nutrition. I like to put away a few pumpkins when they are in season to give to the chickens through the winter.

Storing Pumpkin in the Freezer

Pumpkin will last a long time in a cool, dry environment such as a root cellar. You can mimic a root cellar's conditions using items such as buried trash cans or coolers. Or, you can cook the pumpkin and store the flesh in the freezer until you need it. Even the guts and seeds from the pumpkin can be stored in the freezer.

INGREDIENTS AND MATERIALS

- Baking sheet
- Foil
- Pumpkin
- Knife and cutting board
- Large spoon

1. Preheat the oven to 325°F. Line a baking sheet with foil.
2. Slice the pumpkin in half. Scoop out all the guts and seeds and store separately in the freezer or feed to the flock right away.
3. Place the two sides of the pumpkin cut-side down on the foil-covered pan.
4. Roast the pumpkin in the oven for 30 to 45 minutes. Test for doneness by sticking a fork into the rind. If the fork penetrates into the flesh easily, the pumpkin is cooked.
5. Remove the pan from the oven and let cool.
6. Once you can touch it safely, remove the outer skin. Cut up the pumpkin flesh and store in freezer bags or containers in the freezer.

Stuffed Pumpkin

- 1 medium-sized pumpkin, cut in half
- 16 oz cooked pumpkin puree (canned is okay, but don't use pumpkin pie filling)
- ½ cup oatmeal
- ½ cup mealworms
- ¼ cup flaxseed
- ¼ cup sunflower seeds
- Handful fresh herbs, chopped (reserve a few sprigs for garnish, if desired)
- Raisins or any other dry treats (optional)

1. Scoop out the guts and some of the pumpkin flesh. Save the pumpkin shells. Cut up the long strings of guts. Place the goop on a foil lined baking sheet. Roast at 325°F for 20 minutes.
2. Mix all the ingredients together in a large bowl. Pour/scoop into the hollowed out pumpkin halves.
3. Serve to your flock!

Treats for Molting Season

As daylight begins to shorten, it's time for molting season. The hens may even stop laying eggs during the molt, because all the protein intake is going towards feather growth. If you have extra eggs during the spring and summer seasons, you can freeze the extra eggs for the fall season when you most certainly will see lowered egg production. The molt will start with the feathers on the head and work towards the tail. Some chickens have an easier time with the molt and you might not notice the rougher appearance. Other chickens will look like they have shaken all their feathers off at once.

The best thing you can do to support your chickens during molting season is to feed them adequate protein in the form of a high quality layer ration. Look for a ration that has at least 16 percent protein. You could even switch to a meat bird ration at 18 percent protein. Don't overdo the amount of scratch grains during molt, either. The chickens will still be happy to eat the chicken candy, but it will result in lower protein intake and a slower recovery from molt.

Some treats that you may already have around your home or feed room are great for this time of year. Chickens will always run eagerly towards a handful or two of mealworms or dried soldier fly larvae. These are fantastic for protein intake and rarely will you see a chicken turn them down. Even though the molting chickens will look pleadingly at you, there is no need to overdo the treats during the molt. The old adage, everything in moderation, still applies.

Expect new feather growth in your chickens for the next three to eight weeks. Some chickens molt and recover feathers quickly and some take *forever*. Fear not, your flock will eventually be fully feathered again and ready to fluff up for chilly winter nights. Take care when handling your chickens during the molt as the new feather shafts are delicate and can be injured easily.

Corn Cob Treats

Do you have some dried corn on the cob? Make sure it hasn't been treated with pesticides, as many corn-cob decorations have been. These treats don't require any fancy ingredients or materials and will make your chickens very happy!

INGREDIENTS AND MATERIALS
- Dried field corn on the cob
- Peanut butter
- Assortment of mixed grain, layer feed, mealworms, dried fruit bits, seeds
- String

1. Tie the strings around the top of the corn cob.
2. Mix up a bowl of dried seeds, grains, and mealworms. No need to measure—just use what you have. Small pieces of dried berries and other fruit can be added, too.
3. Spread the mixture onto a large flat tray or wax paper.
4. Spread peanut butter on the dried corn cobs.
5. Roll the corn in the dried seed/grain mix
6. Hang the corn cobs where the chickens can reach to enjoy the treat.

Molting Care Flock Block Recipe

This recipe will supplement the protein, minerals, and vitamins needed for healthy feather growth. Chickens that do not have their nutritional needs met now will head into the winter season with sparse feather growth or a low body conditioning score.

INGREDIENTS

- 2½ cups oatmeal
- ½ cup flaxseed
- ½ cup whole oats
- ½ cup raisins
- ¼ cup peanut butter
- ¼ cup coconut oil
- ¼ cup molasses or honey
- ½ cup sunflower seeds
- 1 cup mealworms or grubs

1. Mix the dry ingredients together in a large bowl. Add the peanut butter, honey, or molasses and coconut oil. Stir to mix.
2. Lightly grease a bread pan. Scoop or pour the mixture into the bread pan.
3. Bake in oven at 325°F for 40 minutes.

4. Cool to room temperature. Transfer to the refrigerator to further solidify the block.
5. Use a knife to loosen around the edges. Turn over onto a baking sheet or tray. I cut ours into three pieces so each coop could have one.

Treats for Broody Hens

Brooding takes a toll on the hens. Broody hens sit for hours and hours, only taking small breaks to grab a bite of food and some water. At least once a day, encourage your broody hen to get off the nest. If you have a delicious treat ready for her, the prospect of leaving her eggs behind for a few minutes is more enticing. Keeping the surrounding area peaceful while the hen is eating will also encourage her to drink water, eliminate waste, and even sneak in a dust bath or preen.

While the broody hen is off the nest, enjoying her food, you have an opportunity to candle the eggs, clean the nest box up, and check for any egg breakage. Letting the hen sit on eggs that are not developing is not good. The egg will eventually begin to rot and the smell is terrible. The egg will also break from the pressure, sometimes exploding, which makes a mess and contaminates the developing eggs with bacteria.

A broody hen needs nutrition to keep herself in good condition, but she will eat less during the brooding period. Spoiling her with high protein, vitamin rich foods helps her come through the process stronger. In addition to the regular layer feed, make some treat mixtures full of fresh vegetables, grains, and protein sources.

Egg Bake with Herbs for Hens

Eggs are a good source of protein not only for humans but for chickens, too. They will gobble them right up! The key to feeding your chickens eggs is to make sure you cook the eggs before feeding them back to the chickens. Feeding raw eggs to your chickens gives them the OK to start eating eggs right from the nest box! No one wants that to happen.

One dish I enjoy cooking for the chickens is a large baked egg dish. I start with eggs from our hen house, but use the ones that are slightly irregular or have some sort of appearance that makes me hesitate to give them to my egg customers.

Add in all sorts of healthy foods such as dark leafy greens, oats, chopped garlic, flaxseed, sunflower seeds, and fresh herbs.

Flax seeds will add additional omega-3 fatty acids to the dish. We all need more of this!

Sunflower seeds are high in fat but a great source of protein and B vitamins. Sunflower seeds are also high in manganese, which is useful for strong healthy feathers with a high gloss!

Greens are healthy and fully of antioxidants and vitamins. People often wonder about using spinach with chickens. Spinach is probably okay in very small quantities, but it does contain a substance, oxalic acid, that inhibits calcium absorption in chickens. Even though spinach is high in calcium, it may be counterproductive to feed it to chickens, although adding a splash of apple cider vinegar to their water can counteract the oxalic acid. Sorrel, a common garden green, is also is high in oxalic acid. Greens such as dark leafy lettuce, kale, Swiss chard, and beet greens are other good choices.

Most herbs are good for chickens, but some chickens prefer some kinds more than others. It's probably an individual taste and aroma thing. I tend to stick with the savory herbs for the egg dish and not the mints. Oregano has some natural worming properties and is a good herb to add to the chicken's diet. I also add parsley and rosemary.

The Egg Bake with Herbs recipe is also good for molting season, when an increase in protein intake helps the hens regrow feathers. You can easily double the amounts for a larger flock.

INGREDIENTS
- 6 eggs
- ½ cup oats or oatmeal
- 2 teaspoons chopped garlic
- 2 tablespoons whole flaxseed
- 2 tablespoons black oil sunflower seeds
- 2 handfuls of mixed greens and herbs, such as spinach, romaine lettuce, kale, chard, fresh basil, fresh oregano, and rosemary. Dried herbs can be substituted. Use what you have on hand or from cleaning out the refrigerator of leftover bits and pieces.

1. Grease an 8"x8" baking pan.
2. Preheat the oven to 325°F.
3. Beat the eggs in a large bowl.
4. Add the oats, garlic, flaxseed, and sunflower seeds. Mix.
5. Incorporate the mixed greens, folding into the egg batter.
6. Bake the egg casserole for 20 minutes or until completely cooked in the center.
7. Cool before feeding to the flock.

Other herbs and botanicals you could add are dandelion blossoms and greens, nasturtiums, and marigold blossoms.

Garlic is always a good choice to add as it also has some natural healing properties and promotes good immune system health.

A word about dairy: I don't give our chickens a lot of dairy products because they can cause some stomach issues and potentially lead to diarrhea. However, as with most things in life, a moderate amount is not a bad thing. It ups the calcium level of the recipe and adds flavor.

What Treats Can be Fed to the Chickens?

In addition to a diet of layer feed, it's fun to bring the chickens table scraps, garden goodness, and foraged foods.

Fruits
Watermelon
Blueberries
Strawberries
Raisins

Vegetables
Kale, Romaine lettuce
Finely shredded carrots
Chopped Broccoli
Corn, or left over corn cobs
Pumpkin, cooked or raw
Garlic
Cooked sweet potatoes

Grains
Oats or Oatmeal
Flaxseeds
Sunflower seeds
Cooked Whole
Grain Rice

These lists are not comprehensive. If any food is not tolerated well by your flock, discontinue its use.

Herbs and Botanicals
Oregano, Thyme, Marjoram
Parsley, Lemon Balm,
Sage, Mint, Plantain leaves

Avoid giving chickens these potentially toxic foods: raw potatoes, avocados, tomato plants, eggplant, raw onions.

Final Thoughts

DIY chicken projects are as wide and varied as your imagination. A simple, sturdy structure, built up from the ground a few inches can provide all the shelter your flock needs to remain safe. Adding other structures, activities, and feeding devices, adds to the enjoyment of tending to a flock. Whatever way you want to provide for your chickens is up to you. The sky is the limit when you let your do-it-yourself ingenuity take the reins! Start with a simple, sturdy brooder for the chicks. As they are growing, get the future coop ready. Add accessories as time permits. The basics are shelter, food, water, and protection. Working in extras and the ideas from this book will add to the enjoyment of raising a backyard flock. The chickens will have fun roaming the yard while you create a chicken wonderland of DIY projects, just for them. Think about what your chickens need or would like, look at what materials you have on hand, and get creative!

Acknowledgments

I have been blessed to work with some amazing editors in my short career as a writer and author. To Abigail Ghering, from Skyhorse Publishing: Thank you for the chance to write this book, and for the infinite patience you showed me as I learned the details of writing a book as opposed to an article or blog post. Check out Abigail's book *The Homesteading Handbook: A Back to Basics Guide to Growing Your Own Food, Canning, Keeping Chickens, Generating Your Own Energy, Crafting, Herbal Medicine, and More* (New York: Skyhorse Publishing, 2011).

My writing career picked up when Pam Freeman, Editor with *Countryside Magazine* and *Backyard Poultry Magazine* hired me to write articles. Pam, you gave me a wonderful opportunity and I am still proud to be writing articles for the magazines. Pam's book is one that should be on every chicken keeper's book shelf. *Backyard Chickens, Beyond the Basics: Lessons for Expanding Your Flock, Understanding Chicken Behavior, Keeping a Rooster, Adjusting for the Seasons, Staying Healthy, and More!* (Minneapolis: Voyageur Press, 2017).

I met Joel Salatin at the Homesteaders of America Conference, while I was writing this book. He's had a huge influence on our approach to homesteading and farming our family land. Mr. Salatin was very encouraging and quickly agreed to write the foreword for this book. His books are well known and have greatly influenced multitudes of new farmers.

Thanks to many supportive friends and peers whose influence and photographs you will see in the pages of this book. Truly these fellow authors, bloggers, and farmers shared not only photographs but insight, stories, and encouragement.

Truly, no work is accomplished well without input and encouragement from friends, family, colleagues, and other sources. From the beginning of this book, family and friends cheered me on. Fellow chicken-keepers offered me photos. My husband worked alongside me to make sure every project looked its best, despite using recycled and salvaged material. His opinion and ideas were invaluable. To my friends, Amy, Brittany, and Kimberly, thank you for doing the tedious task of

proofing the draft. To Jess, thank you my friend, for keeping my website func-
tioning and going strong while I wrote this book. And to my favorite artist, Jacqui
Shreve: thank you for catching my vision for this book and adding the beautiful
artistic details not only to the projects, but also to the pages of this book. I am
surrounded by wonderful people in my life and I feel blessed.

Art Credits

All painting and drawings seen in this work were done by artist Jacqui Shreve.

Websites Contributing Photographs:

Adventure Acres Iowa http://adventureacresiowa.com
A Farm Girl in the Making https://afarmgirlinthemaking.com
The Fewell Homestead http://thefewellhomestead.com
Homestead Honey https://homestead-honey.com
Pasture Deficit Disorder https://pasturedeficitdisorder.com
Happy Days Farm http://happy-days-farm.com
Common Sense Home https://commonsensehome.com